SRA Real Math

PRACTICE WORKBOOK

Grade 3

Stephen S. Willoughby

•

Carl Bereiter

•

Peter Hilton

•

Joseph H. Rubinstein

•

Joan Moss

•

Jean Pedersen

Columbus, OH

The McGraw-Hill Companies

SRAonline.com

Printed in the United States of America.

Send all inquiries to:
SRA/McGraw-Hill
4400 Easton Commons
Columbus, OH 43219

ISBN 0-07-603737-1

4 5 6 7 8 9 BCH 12 11 10 09 08 07

The McGraw·Hill Companies

Practice Lessons

Practice Lessons

LESSON 1.1

Name ______________________ Date ____________

Counting and Estimating

Find the numbers that are written twice, and write them on your paper. Notice that the numbers are not written in order.

1. Start counting from 7.

   7 9 12 8 11 12 10 13 9 16 14 15 17 ________

2. Start counting from 15.

   22 15 24 17 19 17 21 25 19 18 20 23 16 ________

3. Start counting from 87.

   94 87 96 90 92 88 96 93 89 97 95 91 88 ________

4. Start counting from 242.

   252 251 244 249 242 250 244 248 243 245

   252 246 247 ________

Oh no! Ants came to the picnic!

Look at the picture, and then answer each question.

5. Estimate how many ants are on the picnic table. ________

6. Find the actual number of ants. ________

LESSON 1.2

Name ______________________ Date ____________

Review of Addition and Subtraction Facts

Add or subtract to form fact families.

1. $5 + 4 =$ ______
2. $4 + 5 =$ ______
3. $9 - 4 =$ ______
4. $9 - 5 =$ ______
5. $8 + 9 =$ ______
6. $17 - 9 =$ ______
7. $9 + 8 =$ ______
8. $17 - 8 =$ ______

Complete the fact families.

9. $7 - 5 =$ ______
10. $5 +$ ______ $= 7$
11. $2 +$ ______ $=$ ______
12. ______ $- 2 =$ ______
13. $8 + 6 =$ ______
14. ______ $+$ ______ $= 14$
15. $14 -$ ______ $= 8$
16. ______ $- 8 =$ ______

To play this game, point to two cards without looking. To win, the numbers on the two cards must add to 12.

5	4	7	6	8

17. If the first card picked has the number 4, what number must the second card have to win?

    ______

18. If the first card picked has the number 7, what number must the second card have to win?

    ______

LESSON 1.3

Name ______________________ Date __________

Basic Facts and Missing Terms

Find the missing terms.

1. $6 + \_\_\_\_\_\_ = 12$
2. $\_\_\_\_\_\_ + 7 = 16$
3. $\_\_\_\_\_\_ + 5 = 15$
4. $11 - \_\_\_\_\_\_ = 5$
5. $\_\_\_\_\_\_ - 7 = 7$
6. $\_\_\_\_\_\_ - 5 = 3$

Match each word problem with a correct math sentence. Then solve the problem.

7. Tamar checked out 5 books from the library. Her brother also checked out some books, and altogether they had 12 books. How many books did Tamar's brother check out? ________ books

8. Tom and Brad were the only players who scored runs in the baseball game. Tom scored 3 runs and Brad scored 2. How many total runs were scored? ________ runs

9. Natalie had 15 tokens when she went to the arcade. She came home with 4 tokens. How many tokens did Natalie use? ________ tokens

10. Neela had 20 baseball cards, and then she gave 9 to Morris. How many did she have left? ________ cards

a. $20 + 9 = \_\_\_\_\_\_$

b. $5 + \_\_\_\_\_\_ = 12$

c. $15 - \_\_\_\_\_\_ = 4$

d. $3 + \_\_\_\_\_\_ = 2$

e. $3 + 2 = \_\_\_\_\_\_$

f. $20 - 9 = \_\_\_\_\_\_$

g. $15 + 4 = \_\_\_\_\_\_$

h. $5 + 12 = \_\_\_\_\_\_$

LESSON 1.4

Name ____________________ Date ____________

Equalities and Inequalities

Draw the sign to make the statement correct. Use $<$, $=$, or $>$.

1. 15 __________ 52
2. 112 __________ 121
3. 560 __________ 560
4. 1,000 __________ 989
5. 1,616 __________ 1,661
6. 9,540 __________ 9,540
7. 1,400 __________ 4,100
8. 5,250 __________ 2,550
9. 2,468 __________ 2,468
10. 7,351 __________ 7,315

Use $<$, $>$, or $=$ to make the statement correct.

11. $73 + 39$ ______ $37 + 39$
12. $17 + 45$ ______ $15 + 47$
13. $118 + 20$ ______ $180 + 20$
14. $195 - 163$ ______ $190 - 163$
15. $55 - 33$ ______ $50 - 30$
16. $5 + 369$ ______ $369 + 5$
17. $6,899 + 20$ ______ $6,899 + 200$
18. $6,899 - 20$ ______ $6,899 - 200$
19. $55 + 15$ ______ $65 + 16$
20. $55 - 15$ ______ $65 - 25$
21. $200 - 100$ ______ $200 - 150$
22. $839 + 10$ ______ $389 + 10$

LESSON 1.5

Name ______________________ Date ____________

Simple Function Machines

Find the ***in values***, the ***out values***, or the ***rule***.

1

IN	Rule	OUT
5	+4	
9	+4	
	+4	10
14	+4	

2

IN	Rule	OUT
25	−7	
	−7	20
15	−7	
	−7	11

3

IN	Rule	OUT
14		7
12		
10		3
		1

4

IN	Rule	OUT
25		
		37
15		25
		28

Solve.

5 Mrs. Robinson made a function machine for the class to use. When Ben put a card with the number 5 on it into the function machine, a card with the number 7 came out. What is the rule for the function machine? ____________

Name ______________________ Date __________

Place Value

The chart shows the number of hundred thousands, ten thousands, one thousands, hundreds, tens, and ones in each number.

Write the number on your paper.

	Hundred thousands	Ten thousands	One thousands	Hundreds	Tens	Ones	
1	0	1	5	7	2	4	
2	0	0	3	2	0	0	
3	5	0	0	2	4	7	
4	0	6	5	9	2	3	
5	1	0	7	0	6	0	

Answer each question.

6. What does the 7 stand for in the number 567,809? __________
7. Which digit in the number 59,286 stands for the greatest number?

____

Write each number in expanded form.

8. 79,302 ______________________
9. 11,238 ______________________
10. 629,100 ______________________

Solve.

11. Use the digits 7, 9, and 8 once. What is the greatest number you could make? __________
12. How many numbers can you make using each of the digits 3, 8, and 9 once? What are they?

____

LESSON 1.7

Name ______________________ Date ____________

Numbers to 10,000

Fill in the missing numbers. Count up.

1. 1,765; ________ 1,767; ________; ________; 1,770; ________; 1,772

2. ________; 8,808; 8,809; ________; ________; 8,812; ________; 8,814

Fill in the missing numbers. Count down.

3. 4,457; ________; 4,455; 4,454; ________; 4,452; ________; 4,550

4. 9,501; ________; ________; 9,498; ________; ________; 9,495; 9,494

Kerry, Brad, Pedro, and Lin walked around the pond in the park. The distance around the pond is 880 meters.

Complete the table to show how far each student walked.

Student	Distance
Kerry	1,760 meters
Brad	
Pedro	
Lin	

- Kerry walked around the pond twice.
- Brad walked around the pond once and another 200 meters.
- Pedro walked 4 times around the pond and another 50 meters.
- Lin walked around the pond once and another 75 meters.

Answer each question.

5. Who walked the shortest distance? ________

6. Who walked farther than Kerry? ________

7. If another student walked the same distance as Brad and Kerry together, how far did that student walk?

____

LESSON 1.8

Name ______________________ Date ____________

Rounding

Round to the nearest 10.

1. 473 ________
2. 91 ________
3. 889 ________
4. 124 ________
5. 87 ________
6. 18 ________
7. 639 ________
8. 441 ________

Round to the nearest 100.

9. 134 ________
10. 899 ________
11. 321 ________
12. 451 ________
13. 647 ________
14. 289 ________
15. 935 ________
16. 709 ________

Round to the nearest 1,000.

17. 4,323 ________
18. 5,220 ________
19. 8,345 ________
20. 6,677 ________
21. 7,880 ________
22. 1,654 ________
23. 9,120 ________
24. 3,654 ________

LESSON 1.9

Name ______________________ Date ____________

Counting beyond One Million

Use the number 786,409,135 to answer each question.

1. The 7 stands for ______________________
2. The 6 stands for ______________________
3. The 0 stands for ______________________
4. The 1 stands for ______________________
5. The 5 stands for ______________________

Write each number in expanded form.

6. 5,677 ______________________
7. 8,900,067 ______________________
8. 392,060 ______________________

Fill in the missing digits in each number below.

9. Eight million, one hundred sixty-eight thousand, four hundred five

   8, 1☐☐, 4☐5

10. Eleven million, nine hundred seventy-seven thousand, three hundred sixty

    1☐, ☐77, ☐☐☐

11. Fifty-two million, eighty

    ☐2, ☐☐☐, 0☐0

12. Seventy million, seven hundred eighty-seven

    ☐0, 0☐☐, 7☐☐

LESSON 2.1

Name ______________________ Date __________

Regrouping for Addition

Write the standard name for each of these. You may use sticks or play money to help.

1. 0 tens and 13 = ________
2. 2 tens and 1 = ________
3. 2 tens and 11 = ________
4. 5 tens and 16 = ________
5. 8 tens and 17 = ________
6. 6 tens and 12 = ________
7. 7 tens and 10 = ________
8. 5 tens and 19 = ________
9. 1 ten and 15 = ________
10. 12 tens = ________
11. 14 tens and 9 = ________
12. 16 tens and 11 = ________
13. 0 hundreds and 13 tens = ________
14. 2 hundreds and 1 ten = ________
15. 2 hundreds and 11 tens = ________
16. 4 hundreds and 16 tens = ________
17. 8 hundreds and 17 tens = ________
18. 6 hundreds and 12 tens = ________
19. 7 hundreds and 10 tens = ________
20. 5 hundreds and 19 tens = ________
21. 1 hundred and 15 tens = ________
22. 12 hundreds = ________
23. 14 hundreds and 9 tens = ________
24. 16 hundreds and 11 tens = ________

LESSON 2.2

Name ______________________ Date ____________

Adding Two-Digit Numbers

Add. You can use sticks or other materials to help.

1. 25 + 19 = ____
2. 25 + 36 = ____
3. 47 + 46 = ____
4. 52 + 29 = ____
5. 19 + 17 = ____
6. 27 + 34 = ____
7. 67 + 16 = ____
8. 29 + 35 = ____
9. 17 + 43 = ____
10. 26 + 34 = ____
11. 72 + 19 = ____
12. 35 + 35 = ____
13. 43 + 25 = ____
14. 28 + 34 = ____
15. 37 + 37 = ____
16. 71 + 16 = ____

Solve. Show your work.

17. Stephen collected 37 cans. Then he collected 48 more. How many did he collect in all? ________

18. David has 55 stamps. Jonathan has 17 stamps. How many stamps do David and Jonathan have altogether? ________

19. Frederick is 49 inches tall. His older brother Bill is 7 inches taller. How many inches tall is Bill? ________

LESSON 2.3

Name ______________________ Date ____________

Regrouping for Subtraction

Regroup tens and ones. Then rewrite each number to show at least 10 ones and no more than 19 ones.

1. 34 = __________ tens and __________ ones
2. 38 = __________ tens and __________ ones
3. 40 = __________ tens and __________ ones
4. 50 = __________ tens and __________ ones

Regroup hundreds and tens. Then rewrite each number to show at least 1 hundreds and no more than 19 tens.

5. 240 = __________ hundreds and __________ tens
6. 290 = __________ hundreds and __________ tens
7. 300 = __________ hundreds and __________ tens
8. 330 = __________ hundreds and __________ tens

Solve. Show your work.

Sarah has four $10 bills and seven $1 bills. She must pay Alfonso $19.

9. How much money does Sarah have? __________
10. How can Sarah pay $19? ______________________________

____

11. How much will Sarah have left after paying Alfonso? __________

LESSON 2.4

Name ______________________ Date ____________

Subtracting Two-Digit Numbers

Subtract. Use shortcuts if you can.

1. $95 - 89$
2. $32 - 17$
3. $56 - 36$
4. $92 - 24$
5. $53 - 16$
6. $36 - 17$
7. $51 - 36$
8. $92 - 26$
9. $53 - 19$
10. $32 - 15$
11. $53 - 39$
12. $92 - 29$
13. $46 - 37$
14. $31 - 16$
15. $56 - 37$
16. $92 - 35$

Solve. Show your work.

17. Stephanie had 47 marbles. She gave away 29 marbles. How many did she have left? ____________

18. There were 22 apples on a tree. Then a storm came and blew some down. Only 15 apples were left. How many fell? ____________

19. There were 31 roses in Edward's garden. Then he picked some. Now there are 17 roses in his garden. How many did he pick? ____________

LESSON 2.5

Name ______________________ Date ____________

Applications of Addition and Subtraction

Solve. **Show your work.**

1. There are 29 children in Mrs. Moore's class, and 12 are girls. How many boys are in Mrs. Moore's class?

   ____

2. There are 2 third-grade classes. In one class there are 27 students. In the other class there are 29 students. How many students are there altogether in the 2 classes? ____________

3. Nancy has 63¢, and Vera has 29¢.

   a. How much money do they have altogether?

   ____

   b. How much more money does Nancy have?

   ____

Ruben rode his bike 15 blocks east from his home. He then turned around and rode 9 blocks west.

4. How many blocks is Ruben from home?

   ____

5. How many blocks did Ruben ride altogether?

   ____

LESSON 2.6

Name ______________________ Date ____________

Applications with Money

Write how much. Use play money if you need to.

1. 3 one dollar bills
   4 dimes
   1 nickel
   ____

2. 1 five dollar bill
   1 nickel
   ____

3. 4 one dollar bills
   2 quarters
   4 dimes
   7 pennies
   ____

4. 1 twenty dollar bill
   1 five dollar bill
   3 quarters
   1 nickel
   ____

5. 1 ten dollar bill
   3 five dollar bills
   6 dimes
   ____

6. 2 ten dollar bills
   7 one dollar bills
   2 dimes
   4 pennies
   ____

7. 6 one dollar bills
   5 nickels
   7 pennies
   ____

8. 1 five dollar bill
   1 one dollar bill
   1 quarter
   1 nickel
   2 pennies
   ____

LESSON 2.7

Name ______________________ Date __________

Adding Three-Digit Numbers

Add. Use shortcuts if you can.

1. $556 + 238$
2. $125 + 125$
3. $713 + 225$
4. $721 + 103$
5. $325 + 219$
6. $126 + 125$
7. $716 + 225$
8. $230 + 489$
9. $617 + 257$
10. $128 + 124$
11. $835 + 216$
12. $308 + 99$
13. $325 + 325$
14. $122 + 124$
15. $221 + 492$
16. $689 + 302$

Solve. Show your work.

17. Megan ran 320 yards and then stopped to rest. Then she ran 295 more yards. How far did Megan run in all? ______________

18. Harold weighs 122 pounds. Anthony weighs 183 pounds. How much do Harold and Anthony weigh altogether? ______________

19. Angelo had 452 trading cards in his collection. Then he got 106 new trading cards. How many does he have now? ______________

LESSON 2.8

Name ______________________ Date ____________

Subtracting Three-Digit Numbers

Subtract.

1. $\begin{array}{r} 715 \\ -\ 329 \\ \hline \end{array}$

2. $\begin{array}{r} 901 \\ -\ 776 \\ \hline \end{array}$

3. $\begin{array}{r} 642 \\ -\ 296 \\ \hline \end{array}$

4. $\begin{array}{r} 316 \\ -\ 247 \\ \hline \end{array}$

5. $\begin{array}{r} 632 \\ -\ 300 \\ \hline \end{array}$

6. $\begin{array}{r} 742 \\ -\ 196 \\ \hline \end{array}$

7. $\begin{array}{r} 488 \\ -\ 326 \\ \hline \end{array}$

8. $\begin{array}{r} 632 \\ -\ 298 \\ \hline \end{array}$

9. $\begin{array}{r} 999 \\ -\ 267 \\ \hline \end{array}$

Add or Subtract. Watch the signs. Use shortcuts if you can.

10. $\begin{array}{r} 201 \\ -\ 175 \\ \hline \end{array}$

11. $\begin{array}{r} 445 \\ -\ 380 \\ \hline \end{array}$

12. $\begin{array}{r} 268 \\ +\ 108 \\ \hline \end{array}$

13. $\begin{array}{r} 553 \\ +\ 188 \\ \hline \end{array}$

14. $\begin{array}{r} 981 \\ -\ 189 \\ \hline \end{array}$

15. $\begin{array}{r} 539 \\ +\ 108 \\ \hline \end{array}$

16. $\begin{array}{r} 741 \\ -\ 188 \\ \hline \end{array}$

17. $\begin{array}{r} 864 \\ -\ 469 \\ \hline \end{array}$

18. $\begin{array}{r} 539 \\ -\ 268 \\ \hline \end{array}$

19. $\begin{array}{r} 620 \\ -\ 392 \\ \hline \end{array}$

20. $\begin{array}{r} 399 \\ +\ 399 \\ \hline \end{array}$

21. $\begin{array}{r} 647 \\ -\ 376 \\ \hline \end{array}$

LESSON 2.9

Name ______________________ Date ____________

Approximation

In each problem, two of the answers are clearly wrong and one is correct. Choose the correct answer.

1. 75 + 150 =
   a. 145
   b. 75
   c. 225

2. 507 + 368 =
   a. 425
   b. 875
   c. 8,015

3. 4,327 + 2,581 =
   a. 6,908
   b. 378
   c. 3,488

4. 1,276 + 2,724 =
   a. 6,000
   b. 2,000
   c. 4,000

5. 9,658 + 324 =
   a. 6,752
   b. 9,982
   c. 2,582

6. 850 − 164 =
   a. 686
   b. 976
   c. 26

7. 1,179 − 536 =
   a. 1,273
   b. 643
   c. 1,173

8. 9,085 − 7,281 =
   a. 374
   b. 8,764
   c. 1,804

9. 732 − 458 =
   a. 64
   b. 274
   c. 1,084

10. 5,468 − 379 =
    a. 2,379
    b. 5,089
    c. 1,589

LESSON 2.10

Name ______________________ Date ____________

Adding with 3 or More Addends

Add. Use shortcuts if you can.

1. $\begin{array}{r} 42 \\ 97 \\ +\ 54 \\ \hline \end{array}$

2. $\begin{array}{r} 150 \\ 150 \\ +\ 150 \\ \hline \end{array}$

3. $\begin{array}{r} 367 \\ 489 \\ +\ 201 \\ \hline \end{array}$

4. $\begin{array}{r} 169 \\ 215 \\ 108 \\ +\ 113 \\ \hline \end{array}$

5. $\begin{array}{r} 22 \\ 46 \\ +\ 79 \\ \hline \end{array}$

6. $\begin{array}{r} 147 \\ 153 \\ +\ 142 \\ \hline \end{array}$

7. $\begin{array}{r} 369 \\ 482 \\ +\ 208 \\ \hline \end{array}$

8. $\begin{array}{r} 166 \\ 212 \\ 105 \\ +\ 110 \\ \hline \end{array}$

9. $\begin{array}{r} 97 \\ 65 \\ +\ 78 \\ \hline \end{array}$

10. $\begin{array}{r} 159 \\ 152 \\ +\ 154 \\ \hline \end{array}$

11. $\begin{array}{r} 362 \\ 481 \\ +\ 205 \\ \hline \end{array}$

12. $\begin{array}{r} 416 \\ 219 \\ 133 \\ +\ 179 \\ \hline \end{array}$

13. $\begin{array}{r} 95 \\ 60 \\ +\ 78 \\ \hline \end{array}$

14. $\begin{array}{r} 146 \\ 149 \\ +\ 152 \\ \hline \end{array}$

15. $\begin{array}{r} 592 \\ 608 \\ +\ 139 \\ \hline \end{array}$

16. $\begin{array}{r} 708 \\ 122 \\ 143 \\ +\ 139 \\ \hline \end{array}$

LESSON 2.11

Name ______________________ Date ____________

Adding and Subtracting Four-Digit Numbers

Add or subtract these four-digit numbers. Watch the signs. Use shortcuts if you can.

1. $6281 + 1549$
2. $2473 + 2887$
3. $2906 + 1369$
4. $9365 - 3868$
5. $4325 - 2572$
6. $6175 - 3746$
7. $3008 - 1898$
8. $8903 - 5639$
9. $4000 - 1249$
10. $6000 + 3816$
11. $7946 + 1164$
12. $4183 + 1909$
13. $2679 + 348$
14. $4183 + 1350$
15. $3887 - 3657$
16. $3752 - 2837$
17. $2438 - 1569$
18. $7235 - 5827$
19. $6579 - 2463$
20. $9352 - 5633$

Name ______________________ Date ____________

Adding and Subtracting Very Large Numbers

Add or subtract. Watch the signs.

1. 283178789 + 672484972

2. 100101110 − 79872438

3. 485929111 − 108082483

4. 844992105 + 484773925

5. 792442698 − 485952899

6. 308536702 − 119648885

7. 720003592 − 15988167

8. 558452824 + 619788392

9. 399999999 + 98798876

10. 221401122 − 198795486

11. 520000142 − 496789328

12. 468789205 − 468788990

13. 466522799 + 297132178

14. 6895844254 − 998326312

15. 160848793 + 108949056

LESSON 2.13

Name ______________________ Date ____________

Solving Problems

Here is the number of baseball cards that each student has.

Leon: 324 Paula: 256 Scott: 437 Robyn: 480

Think about odd and even numbers as you solve the following problems.

1. How many baseball cards do the students have altogether? ____________
2. Who has the most cards? ____________
3. How many more cards does Leon have than Paula?
   ____
4. How many more cards does Robyn have than Scott? ____________
5. If Scott gives 24 cards to Paula, how many cards does she have altogether? ____________

Use the map to solve the following problems.

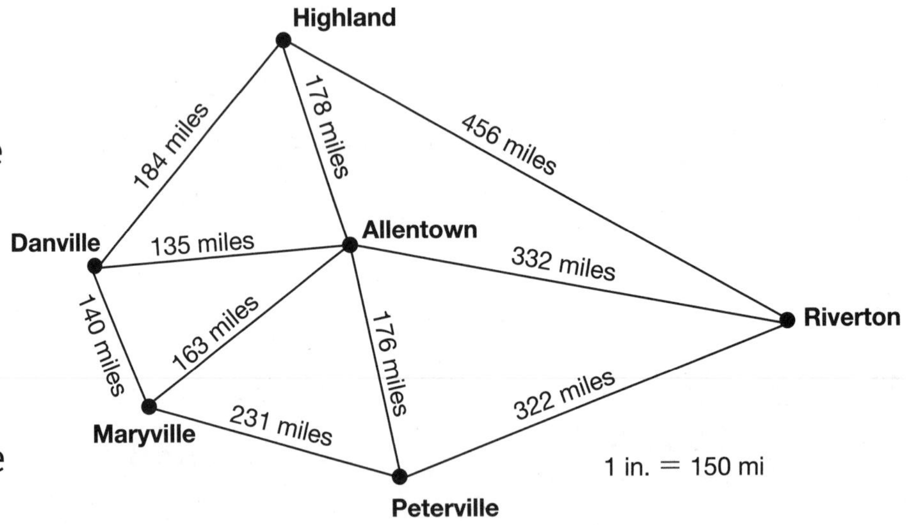

6. What is the distance between Danville and Riverton if you go through Allentown?
   ____
7. What is the distance between Highland and Peterville if you go through Danville and Maryville? ____________
8. What is the shortest route between Maryville and Riverton?
   ____

Name ______________________ Date ____________

Roman Numerals

Write the Arabic numeral for each of these Roman numerals.

I = 1 V = 5 X = 10 L = 50 C = 100

1. IV ________
2. CCCIII ________
3. CXI ________
4. XI ________
5. LXXXIV ________
6. LXXX ________
7. XXX ________
8. CCCXL ________
9. CCCLXII ________
10. XVI ________
11. XLI ________
12. CCLXIII ________
13. XL ________
14. LXVI ________
15. XXXIV ________
16. LXXI ________
17. CLXXXV ________
18. LXX ________
19. XXVIII ________
20. CCXV ________
21. CXLVIII ________
22. LXV ________
23. CCCIX ________
24. CXCVIII ________
25. CCX ________
26. XCV ________
27. CLIX ________
28. CIII ________
29. CIX ________
30. LXXVII ________

Solve. **Show your work.**

31. Allison learned about King Louis XIV in history class. Write the Roman numerals number in Arabic numerals. ________
32. A poem in a book had the Roman numeral XLV above it. What was the number of the poem in Arabic numerals? ________
33. Nero became emperor of Rome in the year 54. How did he write the year in Roman numerals? ________

LESSON 3.1

Name ______________________ Date __________

Telling Time

Tell **the time in three ways.**

1. 08:50 ________, ________ minutes after ________, ________ minutes to ________

2. 02:32 ________, ________ minutes after ________, ________ minutes to ________

3. 11:45 ________, ________ minutes after ________, ________ minutes to ________

4. 06:04 ________, ________ minutes after ________, ________ minutes to ________

5. 12:18 ________, ________ minutes after ________, ________ minutes to ________

Solve.

6. Barry went to the store at 8:15. When he got back home, it was 9:05. How long was he gone? ________

7. It takes Simeon 35 minutes to get home from work. What time should he leave work if he wants to be home by 6:00? ________

Name ______________________ Date ____________

Reading a Thermometer

Write the temperature shown on each Fahrenheit thermometer.

1

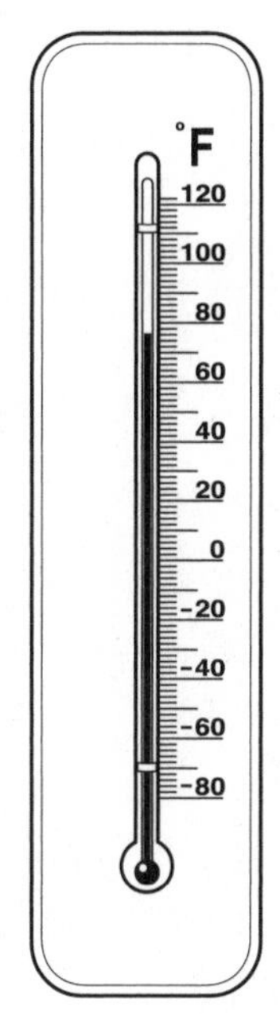

☐ °F

2

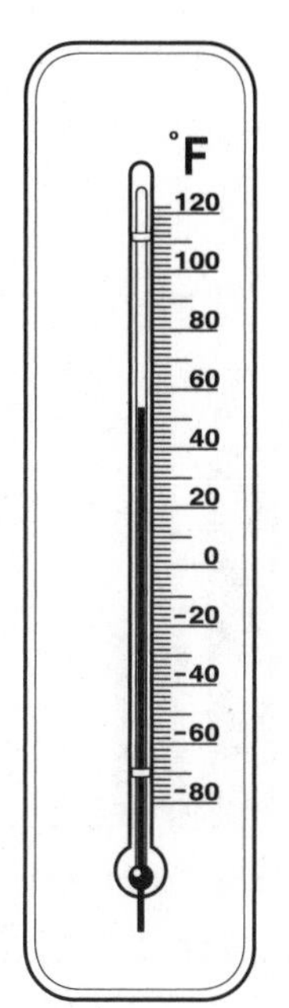

☐ °F

3

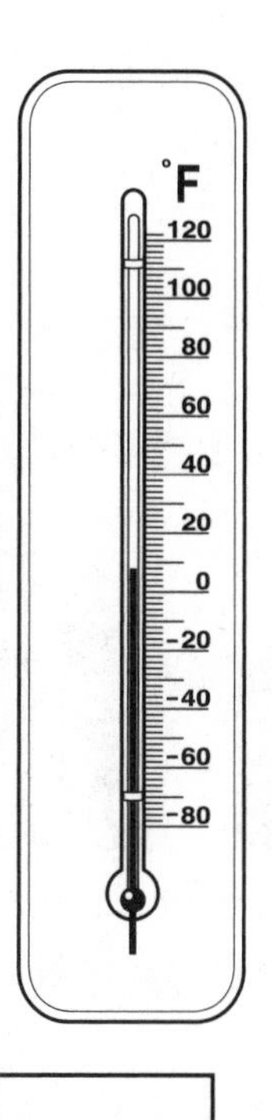

☐ °F

4

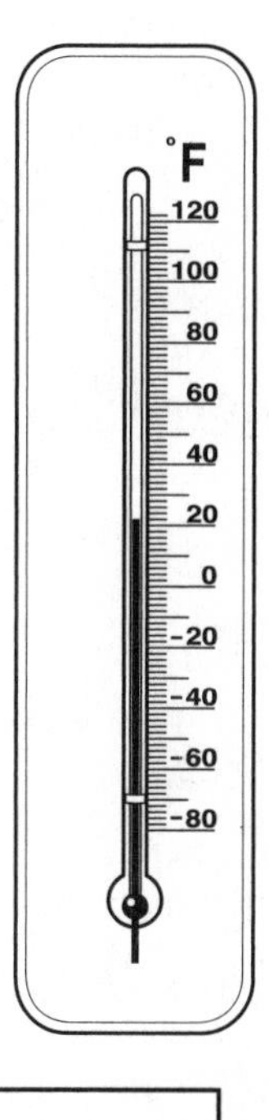

☐ °F

5

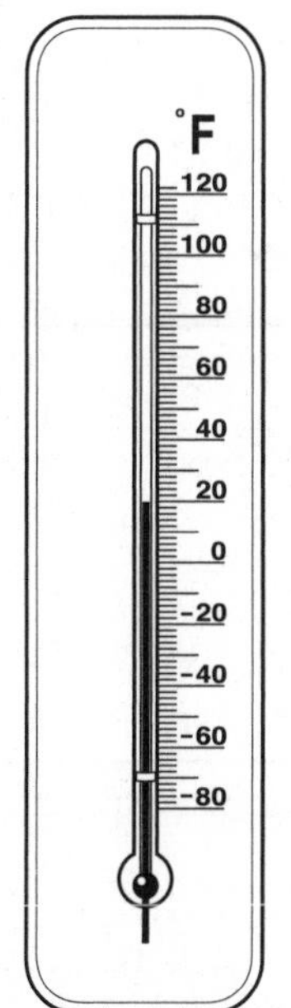

☐ °F

6

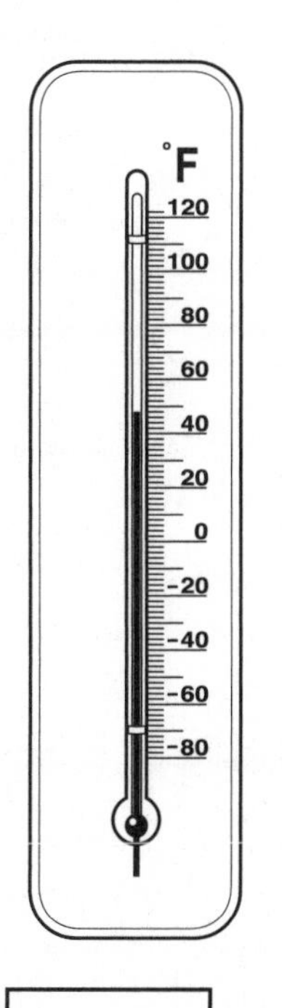

☐ °F

7

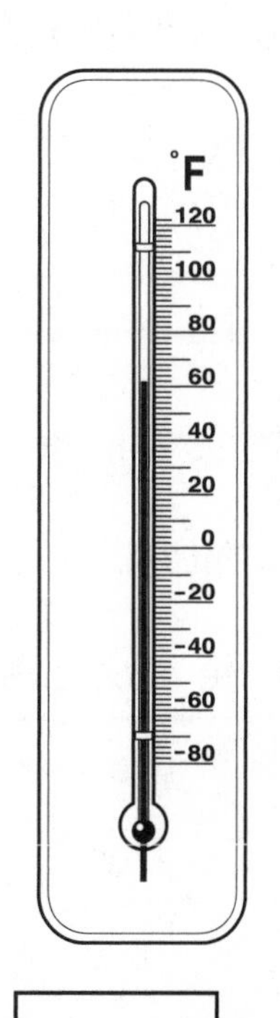

☐ °F

8

°F
120
100
80
60
40
20
0
-20
-40
-60
-80

☐ °F

LESSON 3.3

Name ______________________ Date ____________

Measuring and Graphing

Use the graph to answer these questions.

1. How many children are 45 inches tall? ____________
2. How many more children are 46 inches tall than are 49 inches tall? ____________
3. How many children are in the class? ____________
4. Which height are most of the children? ____________
5. Which heights are the fewest children? ____________
6. How many more children are 48 inches tall than are 42 inches tall? ____________

Name ______________________ Date __________

Measuring Length—Centimeters and Meters

Estimate the length. Then measure to check with your centimeter ruler.

1.

_______ centimeters

2.

_______ centimeters

3. 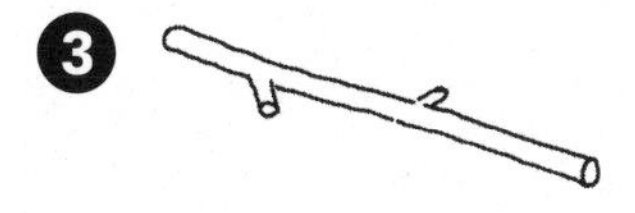

_______ centimeters

4. 

_______ centimeters

Study the picture. Then estimate the answers to these questions. The dog in the middle is about 80 cm tall.

5. About how many centimeters tall is the man?

____

6. About how many centimeters tall is the smaller dog? _______

LESSON 3.5

Name ______________________________ Date ______________

Measuring Length—Inches, Feet, and Yards

Estimate the length. Then measure to check with your inch ruler.

1. 

_________ inches

2. 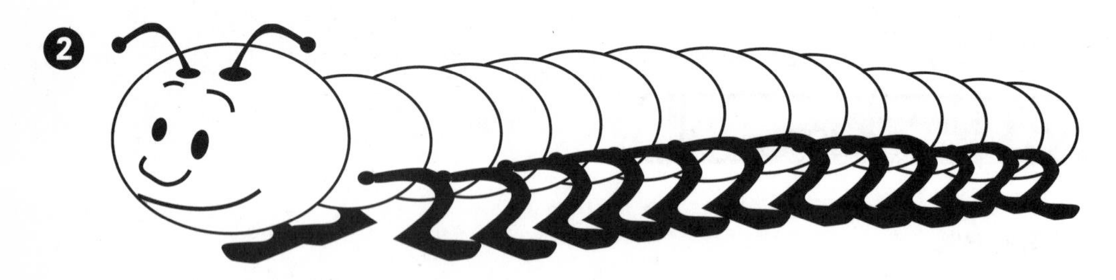

_________ inches

3. 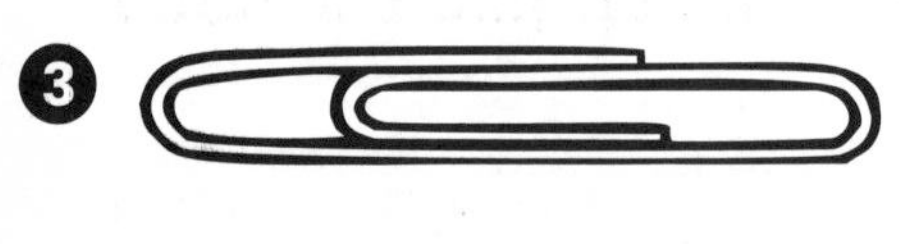

_________ inches

Solve these problems.

A branch 9 feet long fell from a tree in Mr. Axelrod's yard. He broke it into 3 equal pieces to put into a fire.

4. How many feet was each piece? _______________

5. How many yards was each piece? _______________

6. How many inches was each piece? _______________

LESSON 3.6

Name ______________________ Date ____________

Perimeter

Write the name of each polygon. Then find the perimeter.

❶

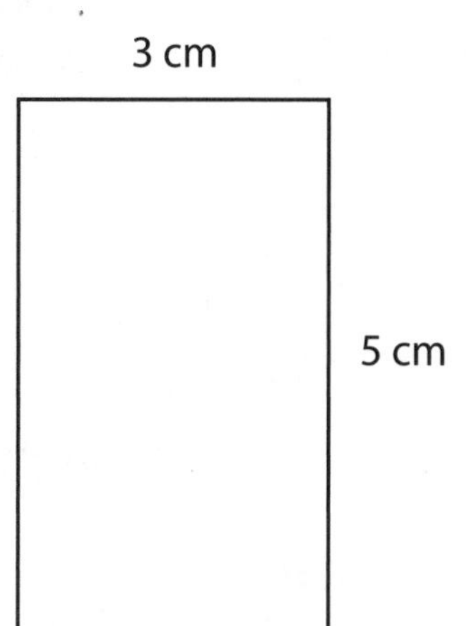

Name: ______________________

Perimeter: __________ centimeters

❷

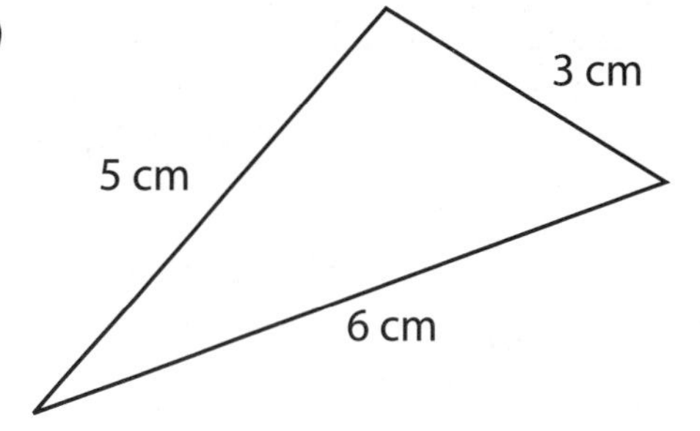

Name: ______________________

Perimeter: __________ centimeters

❸

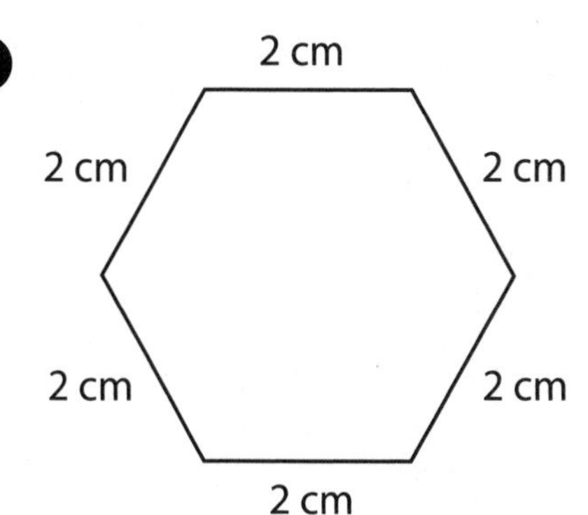

Name: ______________________

Perimeter: __________ centimeters

❹

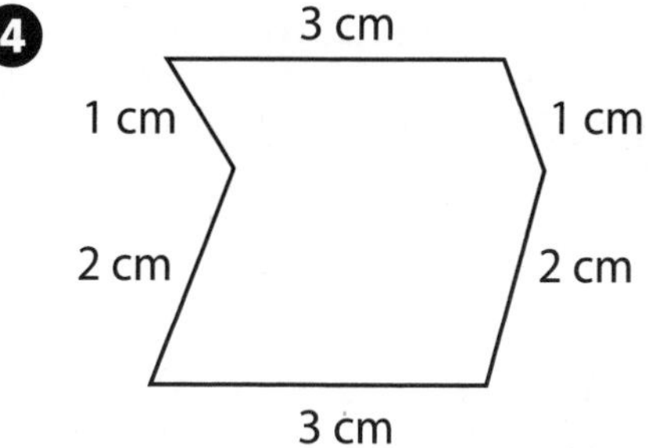

Name: ______________________

Perimeter: __________ centimeters

LESSON 3.7

Name ______________________ Date __________

Perimeter and Area

Find the perimeter and the area for each figure.

1. 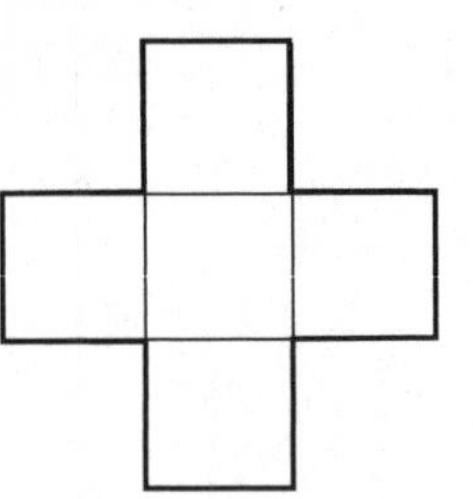

Perimeter: __________ centimeters

Area: __________ square centimeters

2. 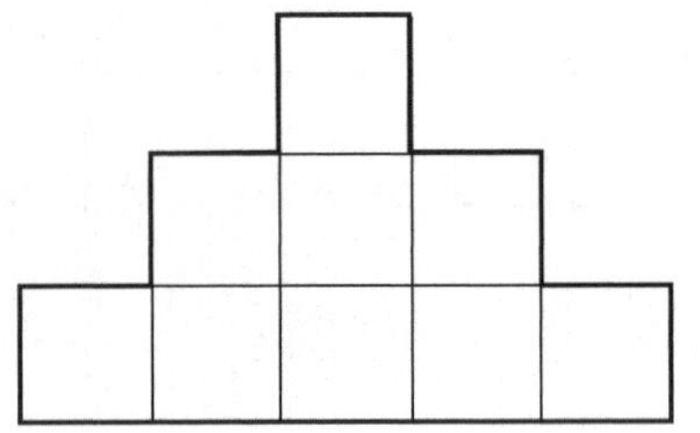

Perimeter: __________ centimeters

Area: __________ square centimeters

3. 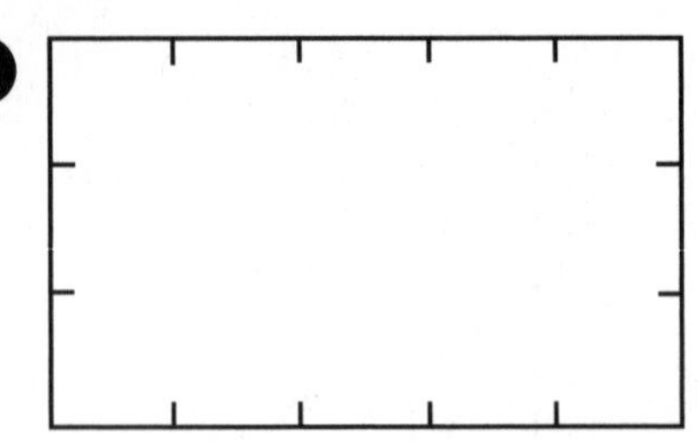

Perimeter: __________ centimeters

Area: __________ square centimeters

4. 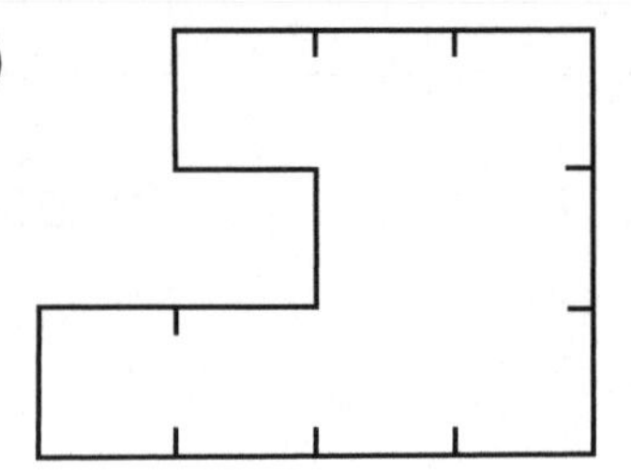

Perimeter: __________ centimeters

Area: __________ square centimeters

Solve these problems.

5. A piece of plywood is in the shape of a triangle. The dimensions are 12 yards, 16 yards, and 15 yards. Find the perimeter. __________

Bill has a piece of carpet in the shape of a rectangle. The dimensions are 10 feet long and 8 feet wide.

6. Find the perimeter of the piece of carpet. __________

7. Find the area of the piece of carpet. __________

LESSON 3.8

Name ______________________ Date __________

Estimating Area

Write the name of each polygon. Then find the perimeter.

1.

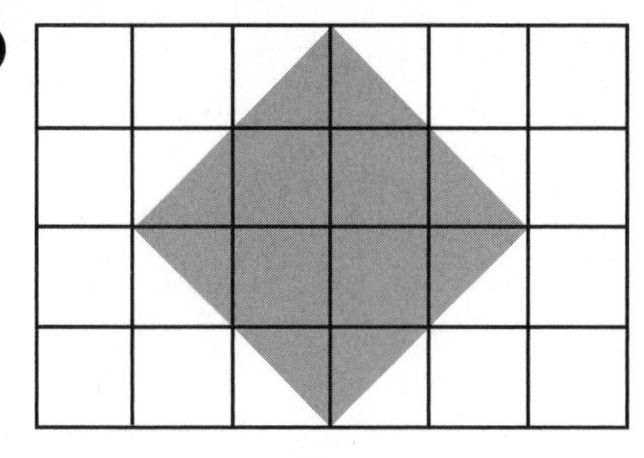

______________ square centimeters

2.

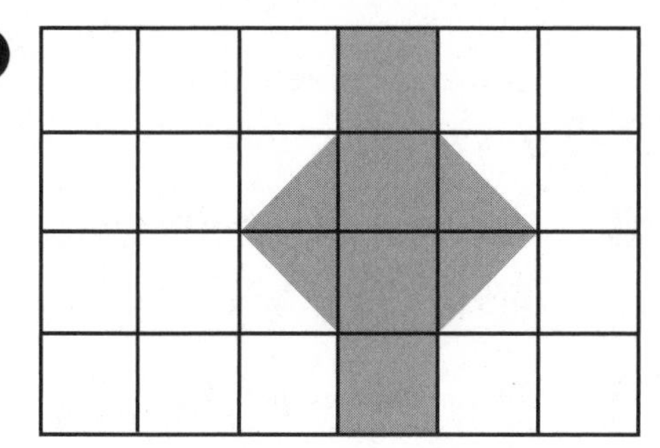

______________ square centimeters

3.

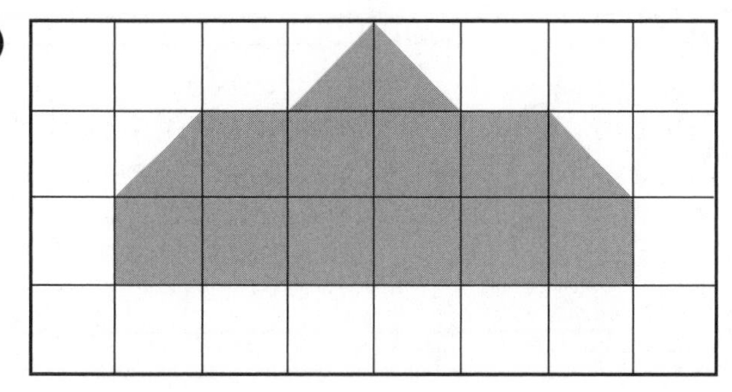

______________ square centimeters

4.

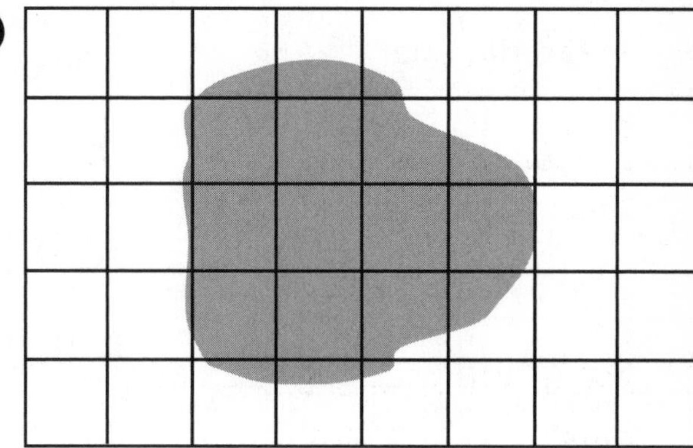

______________ square centimeters

5.

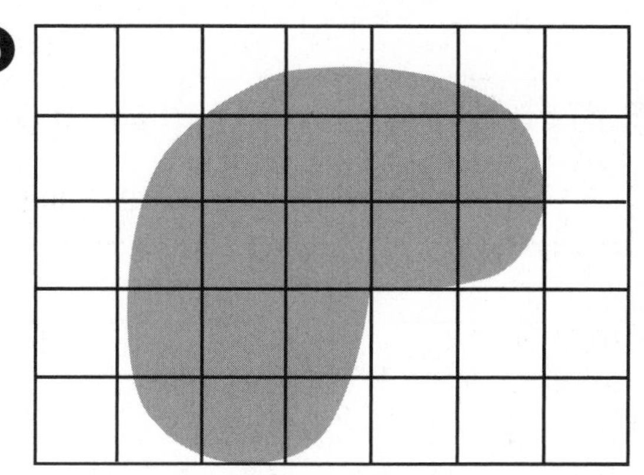

______________ square centimeters

6.

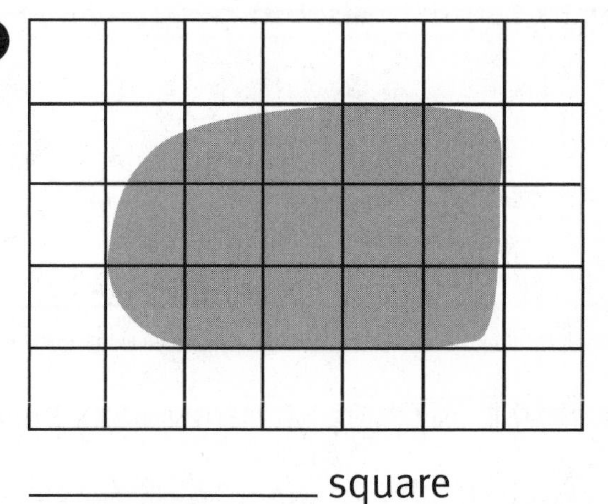

______________ square centimeters

LESSON 3.9

Name ______________________ Date __________

Reading Graphs

Fill in the pictograph below. Red has already been done.

Favorite Color	
Color	**Tally**
red	𝍸 ‖ (7)
blue	𝍸 𝍸 (10)
green	||| (3)
purple	𝍸 ||| (8)
yellow	‖ (2)

Color

red
blue
green
purple
yellow

Number of Students

= 2 students

Use the following bar graph to answer the questions.

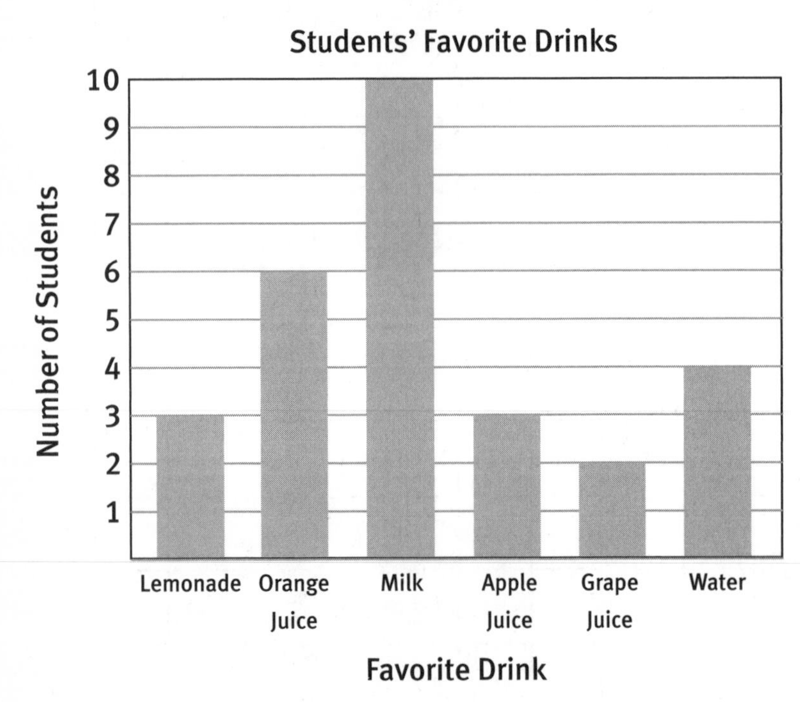

1. How many students were surveyed? __________
2. Which drink did the most students say was their favorite? __________
3. Which drink did the least students say was their favorite? __________
4. How many students said apple juice was their favorite? __________
5. How many more students prefer milk than orange juice? __________
6. What drink is preferred less than lemonade? __________

LESSON 3.10

Name ______________________ Date ____________

Interpreting Bar Graphs

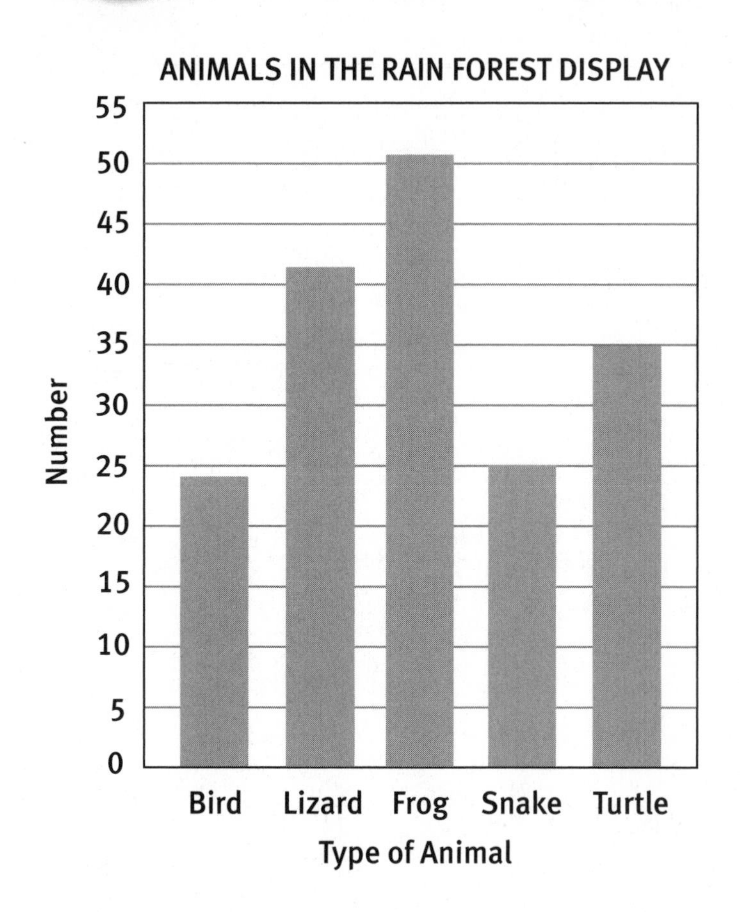

Use the graph to answer the following questions.

1. About how many turtles are in the zoo's rain forest display?

   ____

2. About how many frogs are there?

   ____

3. About how many birds are there?

   ____

4. Lizards, snakes, and turtles are reptiles. About how many reptiles are there?

   ____

5. About how many more lizards are there than snakes?

   ____

6. The zoo added five rare tree frogs to the display. About how many frogs are there now? Write how you know.

   ____

7. Which two types of animal does the display have in about equal numbers? Write how you know.

   ____

LESSON 4.1

Name ______________________ Date ____________

Understanding Multiplication

Solve.

1. There are 6 flowers in each row. How many flowers are there?

$4 \times 6 =$ ________

2. There are 5 flowers in each box. How many flowers are there?

$7 \times 5 =$ ________

3. There are 2 stamps on each letter. How many stamps are there?

$8 \times 2 =$ ________

4. There are 3 milk cartons on each lunch tray. How many milk cartons are there?

$3 \times 3 =$ ________

5. How many books are there?

$2 + 6 + 3 =$ ________

LESSON 4.2

Name ______________________ Date __________

Skip Counting

Use skip counting to find the missing numbers and then to multiply. You may use the pictures to help.

1. 3, 6, ______, 12, ______, 18

2. $2 \times 3 =$ ______
3. $6 \times 3 =$ ______

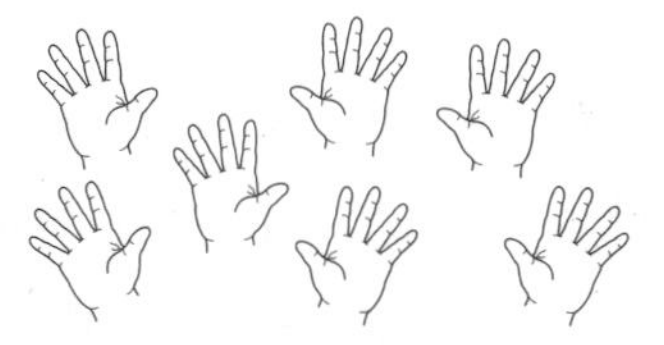

4. 5, 10, 15, ______, ______, ______, 35

5. $4 \times 5 =$ ______
6. $6 \times 5 =$ ______

Find the missing numbers in each skip counting exercise. Use the completed problem to help multiply.

7. 8, 16, 24, ______, 40, ______, ______, 64

8. $3 \times 8 =$ ______
9. $6 \times 8 =$ ______

10. 4, 8, 12, ______, ______, 24, ______, 32

11. $2 \times 4 =$ ______
12. $5 \times 4 =$ ______

Use skip counting to find out the number of rings. Then do the multiplication.

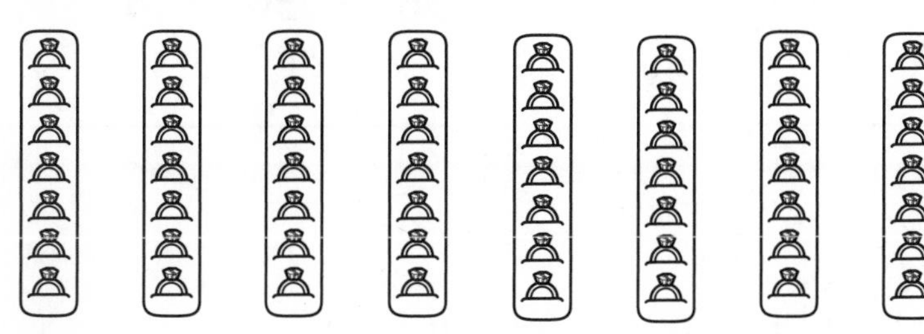

13. 7, 14, ______, ______, 35, 42, ______, ______

14. $8 \times 7 =$ ______

LESSON 4.3

Name ______________________ Date ____________

Multiplication and Number Patterns

Follow the directions for marking numbers on the 100s chart. Then answer the questions.

0	1	2	3	4	5	6	7	8	9
10	11	12	13	14	15	16	17	18	19
20	21	22	23	24	25	26	27	28	29
30	31	32	33	34	35	36	37	38	39
40	41	42	43	44	45	46	47	48	49
50	51	52	53	54	55	56	57	58	59
60	61	62	63	64	65	66	67	68	69
70	71	72	73	74	75	76	77	78	79
80	81	82	83	84	85	86	87	88	89
90	91	92	93	94	95	96	97	98	99

1. Ring the number in the boxes you reach when counting by 3s. Place an *X* in the boxes you reach when counting by 7s.

2. Why are there more numbers with a ring than numbers with an *X?*

   ____

3. How many boxes have both a ring and an *X?* __________

4. What is special about the numbers that have both a ring and an *X?*

   ____

Look at the Venn Diagram below and think about any patterns you may see.

5. What pattern do you see with the numbers in the left circle?

   ____

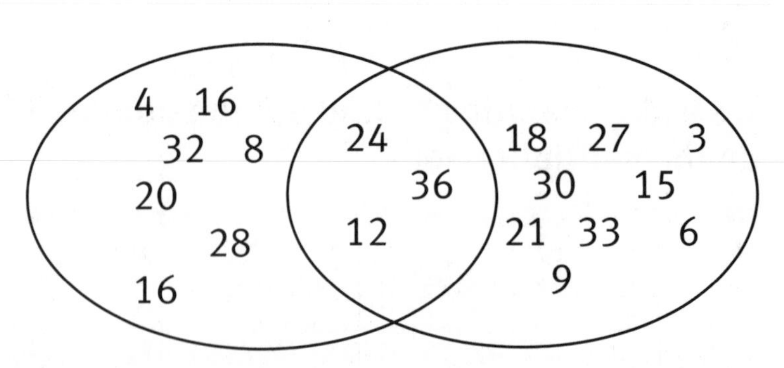

6. What pattern do you see with the numbers in the right circle?

   ____

7. What pattern do you see with the numbers in the middle area?

   ____

LESSON 4.4

Name ________________ Date ________

Multiplication Table

Use the Multiplication Table below to find the answers.

1. $8 \times 5 =$ ______
2. $5 \times 8 =$ ______
3. $6 \times 7 =$ ______
4. $6 \times 10 =$ ______
5. $4 \times 8 =$ ______
6. $8 \times 4 =$ ______
7. $5 \times 9 =$ ______
8. $9 \times 2 =$ ______
9. $2 \times 9 =$ ______

×	0	1	2	3	4	5	6	7	8	9	10
0	0	0	0	0	0	0	0	0	0	0	0
1	0	1	2	3	4	5	6	7	8	9	10
2	0	2	4	6	8	10	12	14	16	18	20
3	0	3	6	9	12	15	18	21	24	27	30
4	0	4	8	12	16	20	24	28	32	36	40
5	0	5	10	15	20	25	30	35	40	45	50
6	0	6	12	18	24	30	36	42	48	54	60
7	0	7	14	21	28	35	42	49	56	63	70
8	0	8	16	24	32	40	48	56	64	72	80
9	0	9	18	27	36	45	54	63	72	81	90
10	0	10	20	30	40	50	60	70	80	90	100

Solve.

10. Each sticker costs 6¢. How much will 7 stickers cost? ______
11. Each bag has 8 peaches. How many peaches are in 9 bags? ______

Ring the correct number sentence, and then solve the problem. Use the multiplication table to help with your computation, if needed.

12. Laura has 9 notebooks. She buys 3 more. How many notebooks does she have now? ______

    a. $9 + 3$ **b.** 9×3

13. Each ride costs \$2. How many dollars will 6 rides cost? ______

    a. $2 + 6$ **b.** 2×6

LESSON 4.5

Name ______________________ Date __________

Arrays

A car dealer has placed the cars for sale in rows.

Answer each question.

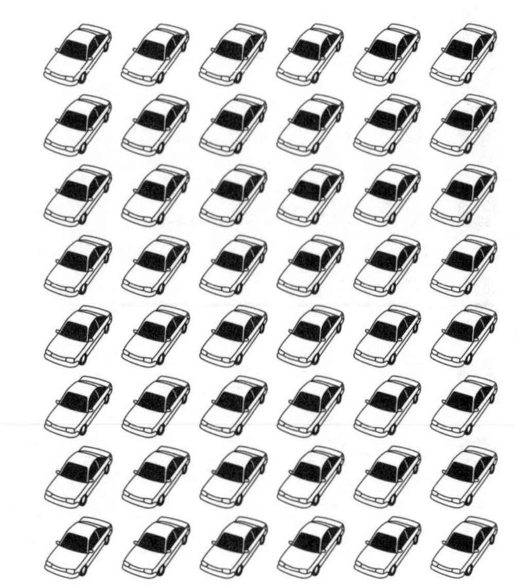

1. How many cars are on the lot? __________

2. Check your answer by skip counting.

   6, 12, ______, 24, 30, 36, ______, ______

3. Check your answer again by skip counting a different way.

   8, 16, 24, 32, ______, ______

4. Write a number sentence to show the number of cars.

   ____

Kim is placing apples on display in his father's fruit store.

5. When Kim is done, there will be seven rows. How many apples will there be in total?

   ____

6. Check your answer by skip counting.

   7, 14, 21, 28, 35, ______, ______

7. Is there another way to skip count to get to 49? ______
   If yes, what number do you use to skip count? ______

8. How many rows of 7 apples does it take to get an even number of apples? __________________

LESSON 4.6

Name ______________________ Date __________

Arrays and Multiplication

Find the area of these shapes.

1. The area is ______ square units. Write a number sentence that shows how you know.

   ____

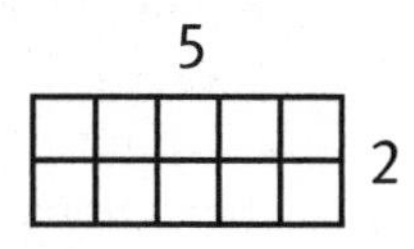

2. The area is ______ square units. Write a number sentence that shows how you know.

   ____

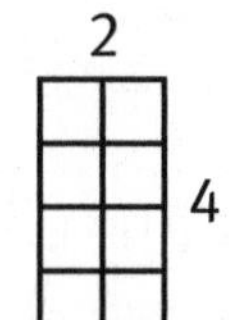

3. The area is ______ square units. Write a number sentence that shows how you know.

   ____

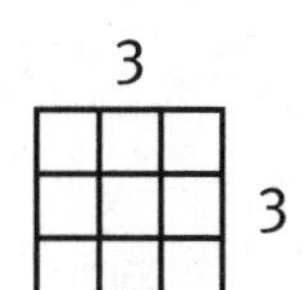

4. The area is ______ square units. Write a number sentence that shows how you know.

   ____

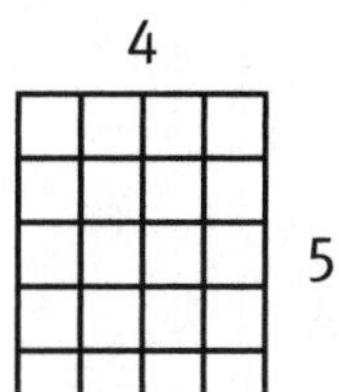

5. The area is ______ square units. What two arrays can you use to get this answer?

   ______ × ______ and ______ × ______

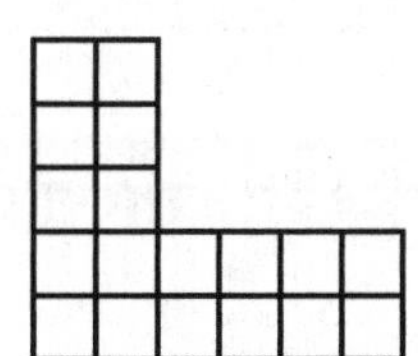

6. The area is ______ square units. What two arrays can you use to get this answer?

   ______ × ______ and ______ × ______

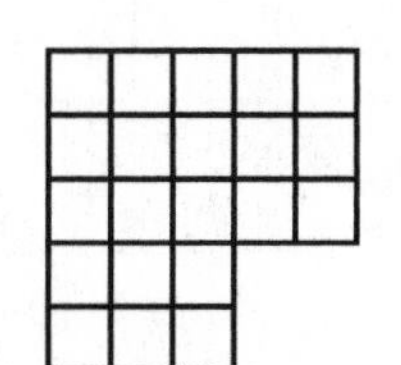

LESSON 4.7

Name ______________________ Date __________

Area and Approximate Measure

Find the area of the shaded part of each rectangle. Remember to use the correct units in your answers.

1. 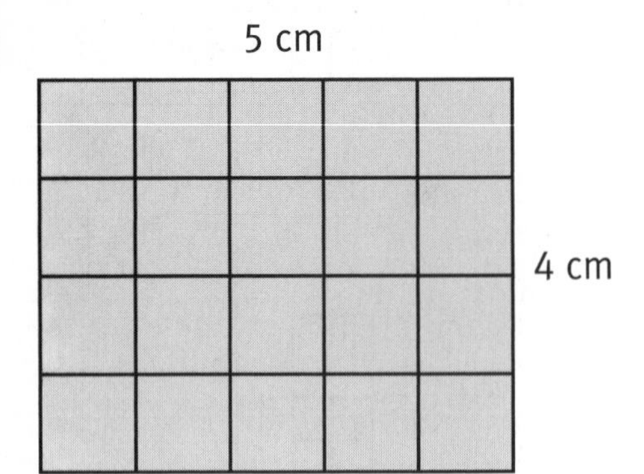

____

2. 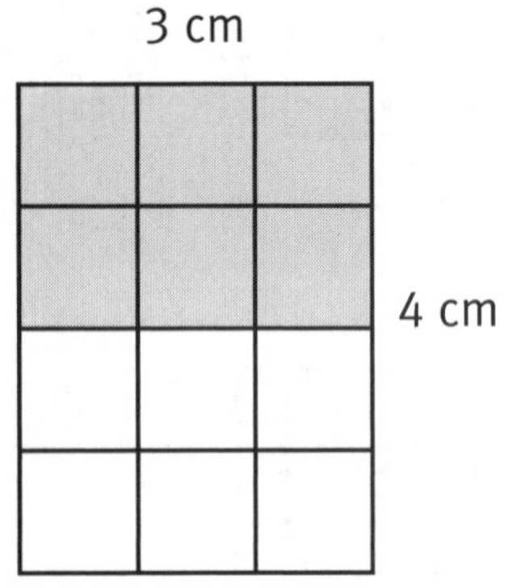

____

3. 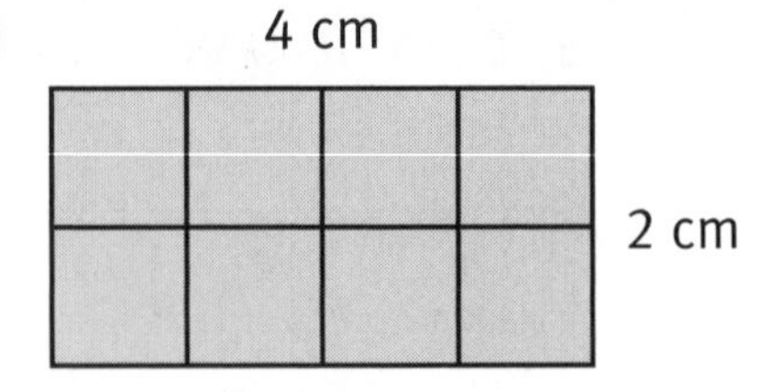

____

4. 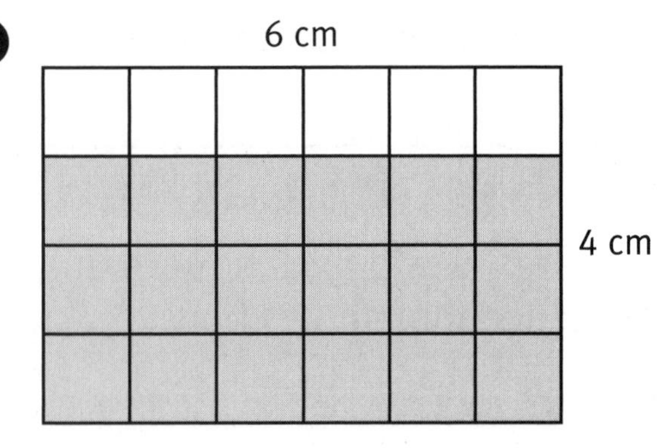

____

Estimate the area of the shaded part of each figure.

5. 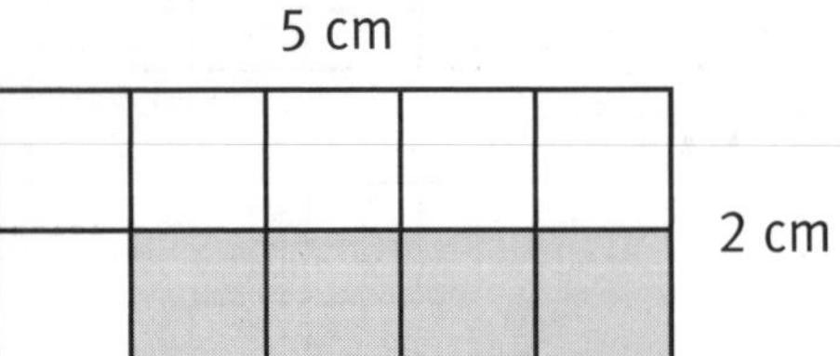

____

6. 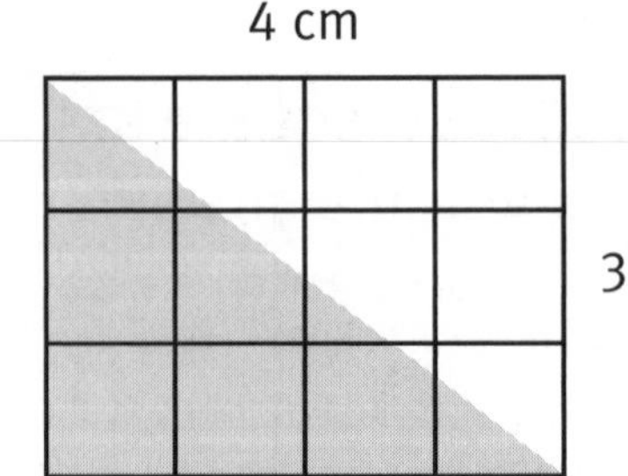

____

LESSON 4.8

Name ____________________ Date __________

Commutative Law of Multiplication

Find the area of the following rectangles.

1. 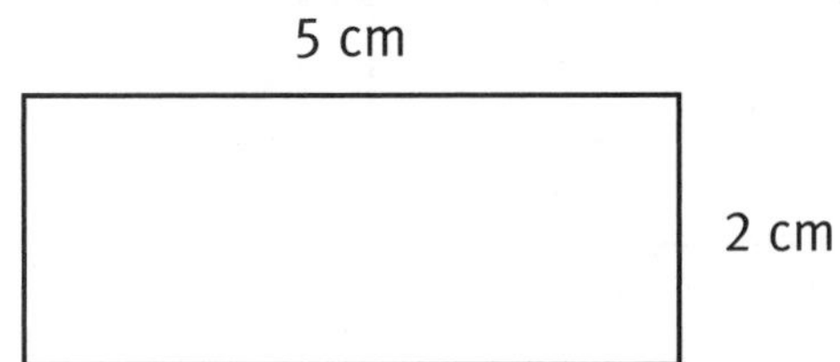

____

3. 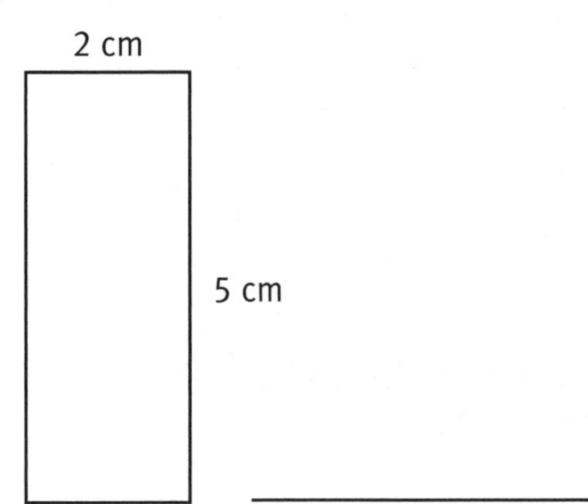

____

2. 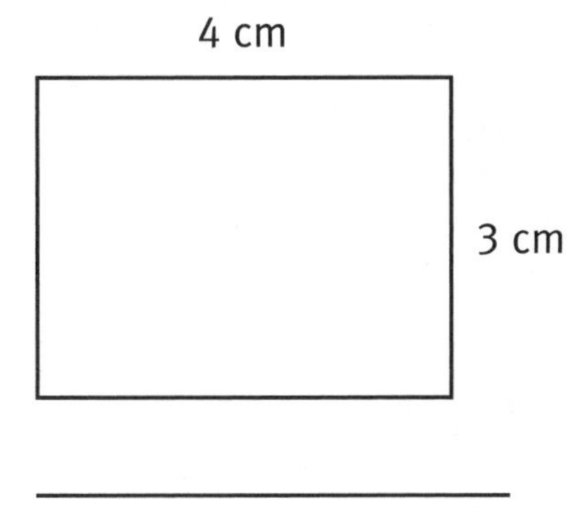

____

4. 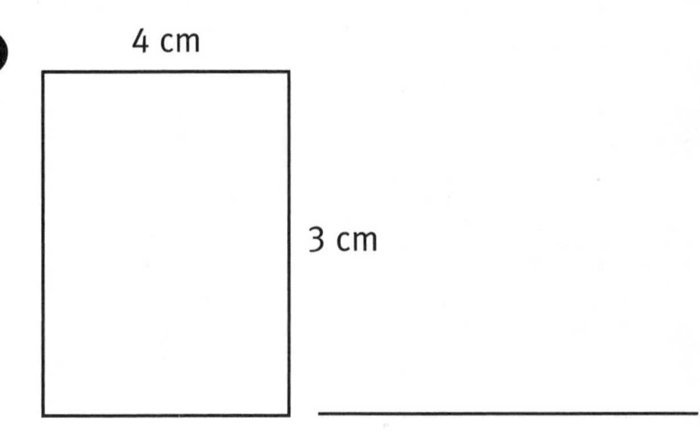

____

Find the product. Use repeated addition, skip counting, or the multiplication table on page 37 to check your answers.

5. $4 \times 6 =$ ___
6. $6 \times 4 =$ ___
7. $2 \times 8 =$ ___
8. $8 \times 2 =$ ___
9. $10 \times 5 =$ ___
10. $5 \times 10 =$ ___
11. $3 \times 9 =$ ___
12. $9 \times 3 =$ ___

Write the multiplication sentence for each problem. Solve.

13. Naomi planted 5 trees each hour. She did this for 3 hours. How many trees did she plant?

    _______ × _______ = _______ trees

14. A small cafe has 8 tables. There are 6 chairs at each table. How many chairs are there?

    _______ × _______ = _______ chairs

LESSON 5.1

Name ______________________ Date ____________

Multiplying by 2, 1, and 0

Use skip counting to fill in the blanks.

1.

28	30			36	38				46

2.

100	102			108		112	

Use skip counting or adding the number to itself to find the products.

3. $2 \times 5 = \square$

4. $2 \times 6 = \square$

5. $7 \times 2 = \square$

6. $2 \times 1 = \square$

7. $2 \times 2 = \square$

8. $2 \times 0 = \square$

Multiply.

9. $1 \times 1 = \square$

10. $2 \times 7 = \square$

11. $1 \times 1 = \square$

12. $0 \times 1 = \square$

13. $1 \times 7 = \square$

14. $2 \times 1 = \square$

15. $0 \times 0 = \square$

16. $0 \times 7 = \square$

17. $2 \times 4 = \square$

Solve each problem. Show your work.

Sarah needs to make 8 gift bags for her birthday party guests.

18. Sarah will put 1 ball into each bag. How many balls will she need? ______

19. Sarah will put 2 stickers into each bag. How many stickers will she need? ______

20. Sarah's mother bought 20 stickers. How many stickers will be left? ______

LESSON 5.2

Name ______________________ Date __________

Multiplying by 10 and 5

Use the patterns for the 5s and 10s facts and other patterns you have learned to find each multiplication fact.

1. $2 \times 10 = \square$
2. $5 \times 5 = \square$
3. $10 \times 6 = \square$
4. $10 \times 5 = \square$
5. $7 \times 2 = \square$
6. $4 \times 5 = \square$
7. $0 \times 5 = \square$
8. $10 \times 8 = \square$
9. $9 \times 10 = \square$

Solve each problem. Show your work.

Matthew and his family raise rabbits. They keep 5 rabbits in each cage. All the cages are full.

10. Could Matthew have 32 rabbits? Explain.

____

11. Could Matthew have 45 rabbits? Explain.

____

12. Matthew has 35 rabbits. How many cages does he have? Explain.

____

Chelsea's mom has a van that holds 10 people. Kyle's dad has a car that holds 5 people.

13. How many trips would Chelsea's mom make to take 40 people to the zoo? __________

14. How many trips would Kyle's dad make to take 35 people to the zoo? __________

LESSON 5.3

Name ______________________ Date ____________

Multiplying by 9

Multiply.

1. $0 \times 9 = \square$
2. $9 \times 1 = \square$
3. $8 \times 9 = \square$
4. $6 \times 9 = \square$
5. $9 \times 7 = \square$
6. $3 \times 9 = \square$
7. Draw 6×9 and 9×6 as arrays.

Describe the array in words (_____ groups of _____) and in a multiplication sentence (_____ × _____ = _____).

8. 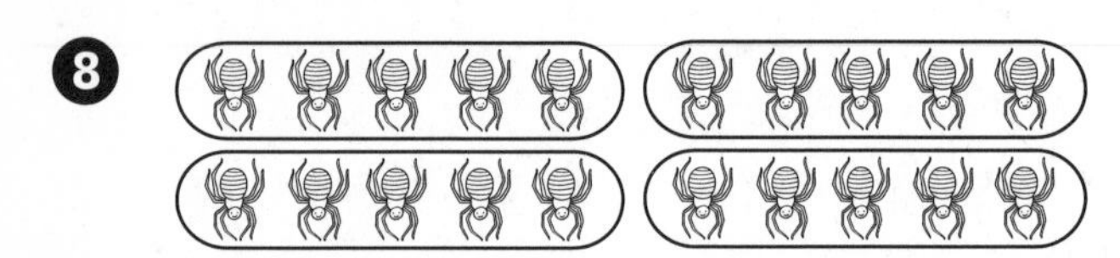

Solve each problem. Show your work.

Tyrone played 6 games of miniature golf this week. Each game had 9 holes.

9. How many holes did Tyrone play? ____________
10. Tyrone played 4 games with Joseph. How many holes did Tyrone play with Joseph? ____________
11. How many holes did Tyrone play when Joseph did not play with him? ____________

Sadie has 36 stuffed bears. She likes to put them in groups around her bedroom.

12. Can Sadie make 4 equal groups of bears? Explain. ____________
13. How many bears will be in each group? ____________
14. Sadie gives 9 bears to her sister. How many does she have left? ____________
15. Sadie now has $9 \times \square$ bears.

LESSON 5.4

Name ______________________ Date __________

Square Facts

Look at each array. Multiply a whole number by itself to find the number of objects in each square. Count the objects to check your answer.

❶ 

____ × ____ = ____ frogs

❷ 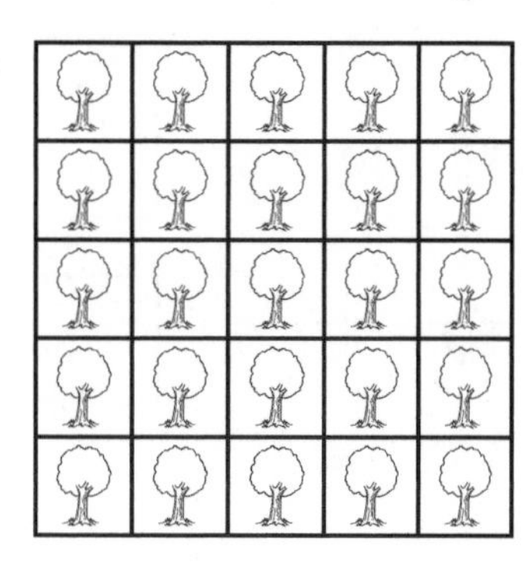

____ × ____ = ____ trees

Multiply.

❸ $1 \times 1 =$ ☐

❹ $4 \times 4 =$ ☐

❺ $7 \times 7 =$ ☐

❻ $2 \times 2 =$ ☐

❼ $5 \times 5 =$ ☐

❽ $8 \times 8 =$ ☐

❾ $3 \times 3 =$ ☐

❿ $6 \times 6 =$ ☐

⓫ $9 \times 9 =$ ☐

Solve each problem.

0	1	2	5	9	10

Choose two numbers from the numbers above. You can use the same number twice. Multiply the two numbers together to answer each question.

⓬ What is the largest answer you can get? Explain. ____________

⓭ What is the smallest number you can get? Explain. ____________

⓮ Which two numbers give an answer of 45? Explain. ____________

⓯ Which two numbers give an answer of 10? Explain. ____________

⓰ Which two numbers give an answer of 18? Explain. ____________

LESSON 5.5

Name ____________________ Date __________

Multiplying by 3 and 6

Multiply.

1. $1 \times 3 =$ ☐
2. $3 \times 7 =$ ☐
3. $6 \times 6 =$ ☐
4. $3 \times 4 =$ ☐
5. $10 \times 6 =$ ☐
6. $6 \times 9 =$ ☐
7. $3 \times 6 =$ ☐
8. $7 \times 6 =$ ☐
9. $6 \times 2 =$ ☐

Solve each problem.

10. Laurie bought 5 balloons. Each balloon costs $3. How much did Laurie pay for the balloons? ____________

11. Laurie's friend bought 9 balloons for $3 each. How much did she pay for the balloons? ____________

12. Tavon gives 6 pencils to each of his friends. He gives pencils to 9 friends. How many pencils does he give away? Explain. ____________

13. ☆☆☆☆☆☆☆
☆☆☆☆☆☆☆
☆☆☆☆☆☆☆
☆☆☆☆☆☆☆
☆☆☆☆☆☆☆

How many stars? ____________

14. A school baseball team is selling bags of nuts at their games. Each bag of nuts costs $3. Copy and complete the table to show how much different numbers of bags will cost.

Bags	1	2	3	4	5	6	7	8	9	10
Cost (Dollars)										

LESSON 5.6

Name ______________________ Date __________

Multiplying by 4 and 8

Multiply.

1. $2 \times 4 = \square$
2. $8 \times 0 = \square$
3. $1 \times 8 = \square$
4. $4 \times 4 = \square$
5. $2 \times 8 = \square$
6. $8 \times 2 = \square$
7. $4 \times 7 = \square$
8. $8 \times 6 = \square$
9. $8 \times 1 = \square$

Solve for *n* in each exercise.

10. $3 \times 4 = n$ $n =$ ____
11. $4 \times 8 = n$ $n =$ ____
12. $6 \times 4 = n$ $n =$ ____
13. $3 \times 8 = n$ $n =$ ____
14. $6 \times 8 = n$ $n =$ ____
15. $9 \times 8 = n$ $n =$ ____

Solve each problem. Show your work.

Mrs. Dresser asked Samuel and Tamika to put 4 ice cubes into each glass.

16. Samuel filled 8 cups with ice. How many ice cubes did he use? ________

17. Tamika filled 5 glasses. How many ice cubes did she use? ________

18. Jamal ordered 6 pizzas. Each pizza cost \$8. How much did Jamal pay for the pizza? ________

LESSON 5.7

Name ______________________ Date __________

Missing Factors and Division I

Solve each problem. Show your work.

1. There are 6 squirrels sharing 24 peanuts. Each squirrel gets the same number of peanuts. How many peanuts does each squirrel get? ______

2. There are 9 mother ducks, and each has the same number of ducklings. There are 36 ducklings. How many ducklings does each mother duck have? ______

Gina is thinking of two numbers. One of the numbers is 6.

3. Suppose the product of the numbers is 54. What is the other number? ______

4. Suppose the product of the numbers is 42. What is the other number? ______

5. There are 32 marbles and 4 children. How many marbles are there for each child?

   $4 \times \square = 32 \quad 32 \div 4 = \square$

6. There are 63 corn plants. There are 7 rows. How many plants are there in each row?

   $7 \times \square = 63 \quad 63 \div 7 = \square$

7. Jamal needs to pay 90¢. He has only dimes. How many dimes will he give to pay? ______

8. Diane has 4 sheets of stickers. Each sheet has 10 stickers. She wants to give the stickers to 5 friends so each friend gets the same number of stickers. How many stickers should each friend get? ______

LESSON 5.8

Name ____________________ Date ____________

Missing Factors and Division II

Fill in the blank.

1. $9 \times 7 = \_\_\_$
2. $\_\_\_ \times 6 = 48$
3. $50 \div 10 = \_\_\_$
4. $56 \div 8 = \_\_\_$
5. $\_\_\_ \div 9 = 7$
6. $48 \div 8 = \_\_\_$
7. $5 \times \_\_\_ = 50$
8. $7 \times \_\_\_ = 56$
9. $\frac{63}{9} = \_\_\_$
10. $\frac{48}{6} = \_\_\_$
11. $\frac{50}{5} = \_\_\_$
12. $\frac{56}{8} = \_\_\_$
13. $\frac{63}{7} = \_\_\_$
14. $\frac{48}{\_\_\_} = 6$
15. $\frac{50}{10} = \_\_\_$
16. $\frac{56}{\_\_\_} = 8$

Solve. Show your work.

17. There are 36 students in the cafeteria. Each table can seat 6 children. How many tables are needed? ____________

18. There are 40 students in the cafeteria. Each table can seat 6 children. How many tables are needed?

____

19. Carla and her friends run a lemonade stand. The money they earn is shared equally. If they earn $12, how much will each person get? ____________

20. Vitali makes 2 pizzas with 6 slices each. He shares the pizzas equally with 3 friends. How many slices does each person get? ____________

LESSON 5.9

Name ______________________ Date __________

Understanding Division

Divide. **Use repeated subtraction if needed.**

1. $30 \div 3 =$ ___
2. $27 \div 9 =$ ___
3. $16 \div 4 =$ ___
4. $49 \div 7 =$ ___
5. $30 \div 5 =$ ___
6. $64 \div 8 =$ ___
7. $16 \div 2 =$ ___
8. $49 \div 1 =$ ___
9. $30 \div 6 =$ ___
10. $36 \div 6 =$ ___
11. $32 \div 8 =$ ___
12. $50 \div 5 =$ ___
13. $30 \div 1 =$ ___
14. $27 \div 3 =$ ___
15. $45 \div 5 =$ ___
16. $16 \div 8 =$ ___

Solve. **Show your work.**

17. Natasha, Chris, April, and Lisa bought 3 packs of pens. Each pack has 8 pens. They divided the pens equally. How many pens did each person get? ______________

18. Beaumont Elementary's third-grade classes are having a field day. Each student must participate in 3 activities. Each class has the same number of students, and a total of 64 students are participating.

    a. How many students are in each third-grade class? Explain your answer. ______________

    b. Students ran 2 at a time in the Relay Race. How many races were there? ______________

    c. There are 8 teams entered in the kickball tournament. How many students are on each team? ______________

Name ______________________ Date ____________

Simple Functions

The tables below show the numbers that went into the number machines. They also show some of the numbers that came out.

Write what you think each function rule is. Then complete the tables.

1 Function Rule: ______

In	Out
9	54
5	30
4	
1	
7	
6	

2 Function Rule: ______

6 in → out 18

In	Out
4	12
3	9
8	
10	
2	
5	

Solve. Show your work.

3 Joe bought 7 bags of balloons.
There are 8 balloons in each bag.
How many balloons did he buy? ____________

4 Jen has 3 baskets of strawberries.
There are 9 strawberries in each basket.
How many strawberries does Jen have? ____________

LESSON 6.2

Name ______________________ Date ____________

Introducing Arrow Notation

The function rule for this function machine is +9.

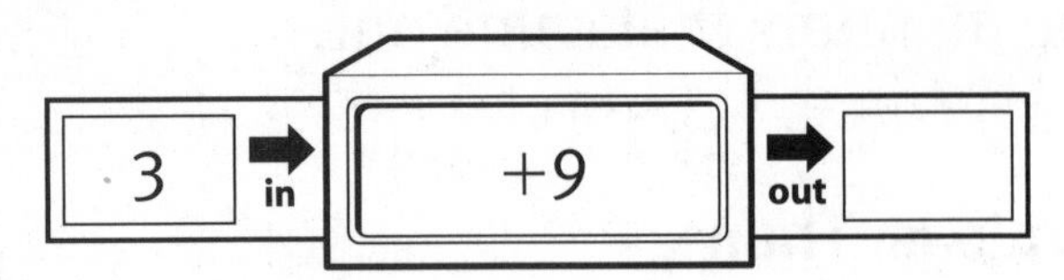

Answer each question and complete each function table.

1. If 3 is put into this function machine, what number will come out?

2. If 5 is put into this function machine, what number will come out?

5 → (+9) → ______

3. If 8 is put into this function machine, what number will come out?

8 → (+9) → ______

Find the function rule for each set of numbers.

4.

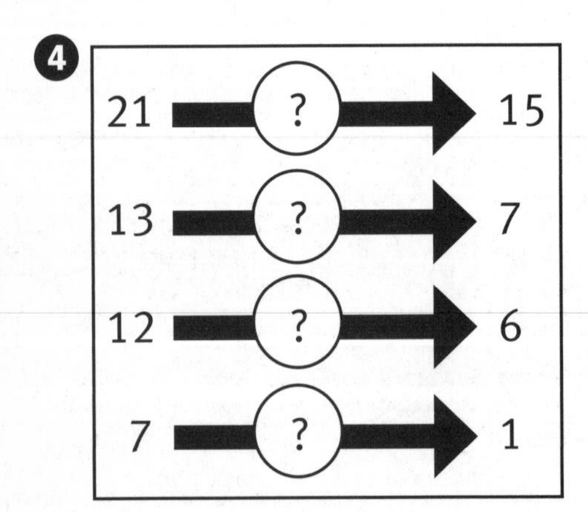

The function rule is ______.

5.

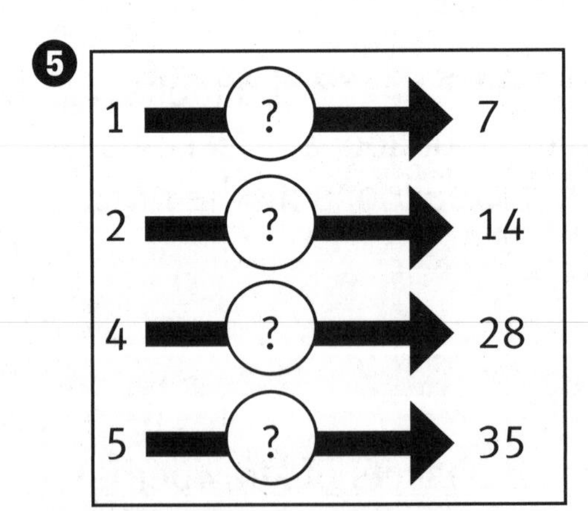

The function rule is ______.

LESSON 6.3

Name ______________________ Date __________

Using Arrow Notation to Solve Equations

Find the values of *n*.

1. $7 \xrightarrow{+8} n$

$n =$ ______

2. $n \xleftarrow{\times 3} 6$

$n =$ ______

3. $n \xleftarrow{\div 4} 28$

$n =$ ______

4. $11 \xrightarrow{-3} n$

$n =$ ______

Find the value of *x*. Then find the value of *y*.

5. $3 \xrightarrow{+1} x \xrightarrow{\times 5} y$

$x =$ ______ $y =$ ______

6. $y \xleftarrow{\times 4} x \xleftarrow{-2} 12$

$x =$ ______ $y =$ ______

7. $2 \xrightarrow{+4} x \xrightarrow{\times 6} y$

$x =$ ______ $y =$ ______

8. $y \xleftarrow{+4} x \xleftarrow{\div 5} 25$

$x =$ ______ $y =$ ______

Find the values of *n*, *x*, and *y*.

9. $n \xrightarrow{+9} 20$

$n =$ ______

10. $9 \xleftarrow{\times 3} x$

$x =$ ______

11. $y \xleftarrow{\times 6} 5$

$y =$ ______

12. $27 \xrightarrow{\div 3} x$

$x =$ ______

13. $15 \xrightarrow{\div 5} n$

$n =$ ______

14. $x \xrightarrow{-10} 11$

$x =$ ______

LESSON 6.4

Name ______________________ Date ____________

Inverse Functions (Reversing the Arrow)

Write the inverse arrow operation.

1. $x \rightarrow (+7) \rightarrow y$ $\quad x \leftarrow (\quad) \leftarrow y$
2. $x \rightarrow (-14) \rightarrow y$ $\quad x \leftarrow (\quad) \leftarrow y$
3. $x \rightarrow (+15) \rightarrow y$ $\quad x \leftarrow (\quad) \leftarrow y$
4. $x \rightarrow (+1) \rightarrow y$ $\quad x \leftarrow (\quad) \leftarrow y$
5. $x \rightarrow (\div 13) \rightarrow y$ $\quad x \leftarrow (\quad) \leftarrow y$
6. $x \rightarrow (\times 8) \rightarrow y$ $\quad x \leftarrow (\quad) \leftarrow y$
7. $x \rightarrow (\div 2) \rightarrow y$ $\quad x \leftarrow (\quad) \leftarrow y$
8. $x \rightarrow (+3) \rightarrow y$ $\quad x \leftarrow (\quad) \leftarrow y$
9. $x \rightarrow (\times 10) \rightarrow y$ $\quad x \leftarrow (\quad) \leftarrow y$
10. $x \rightarrow (-12) \rightarrow y$ $\quad x \leftarrow (\quad) \leftarrow y$

Use inverse operations to find the value of ***n***.

11. $n \rightarrow (\times 8) \rightarrow 16$

    $n =$ ______

12. $n \rightarrow (-9) \rightarrow m \rightarrow (+7) \rightarrow 12$

    $n =$ ______

13. $n \rightarrow (\times 4) \rightarrow m \rightarrow (+3) \rightarrow 19$

    $n =$ ______

14. $11 \rightarrow (-6) \rightarrow m \rightarrow (\div 5) \rightarrow n$

    $n =$ ______

15. $n \rightarrow (\times 5) \rightarrow 25$

    $n =$ ______

16. $n \rightarrow (\div 8) \rightarrow m \rightarrow (\times 2) \rightarrow 6$

    $n =$ ______

17. $9 \rightarrow (-8) \rightarrow n$

    $n =$ ______

18. $28 \rightarrow (\div 2) \rightarrow m \rightarrow (-3) \rightarrow n$

    $n =$ ______

LESSON 6.5

Name ______________________ Date ____________

Variables and Arrow Notation

Solve these problems. Work down the page.

x → (+8) → y

1. If $x = 8$, what is y? ______
2. If $y = 8$, what is x? ______
3. If $x = 10$, what is y? ______
4. If $y = 10$, what is x? ______
5. If $x = 4$, what is y? ______
6. If $y = 4$, what is x? ______
7. If $x = 10$, what is y? ______
8. If $y = 10$, what is x? ______

Complete the function charts.

9. x → (×3) → y

x	y
3	
8	
	3
	30

10. x → (−9) → y

x	y
15	
	20
18	
	1

Ann rides her bike for exercise. It takes her 8 minutes to bike 1 mile. She uses the following equation to find out how long she will need for her ride.

11. On Saturday, Ann wants to bike 3 miles. How long will it take her? ________
12. On Sunday, Ann wants to bike 8 miles. How long will it take her? ________
13. Ann biked for 40 minutes. How many miles did she bike? ________

LESSON 6.6

Name ______________________ Date ______________

Graphing Ordered Pairs

Use the map to answer each question.

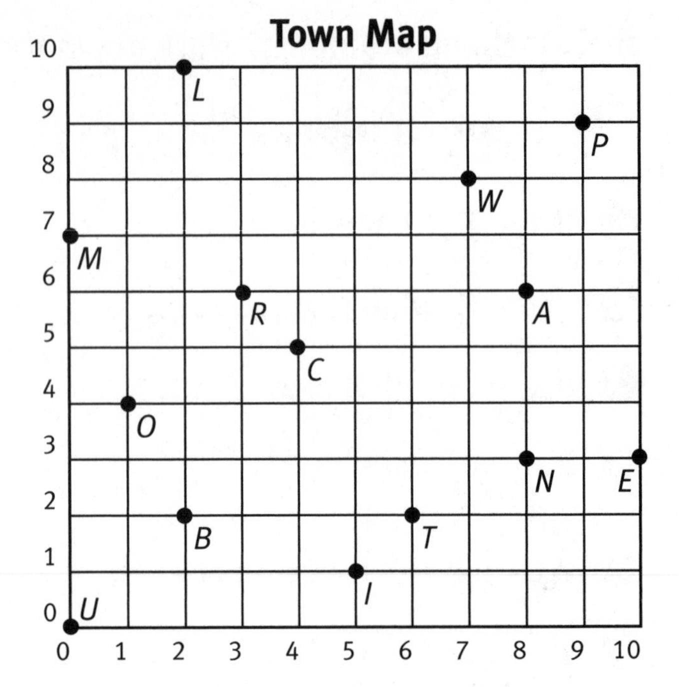

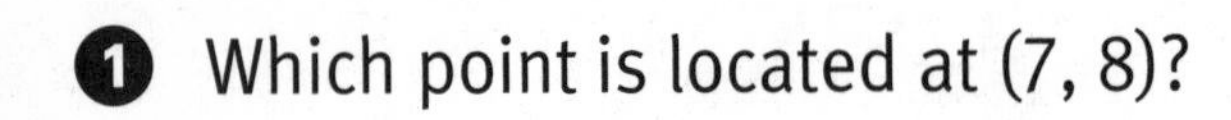

1. Which point is located at (7, 8)?

   ____

2. What is the ordered pair for Point R?

   ____

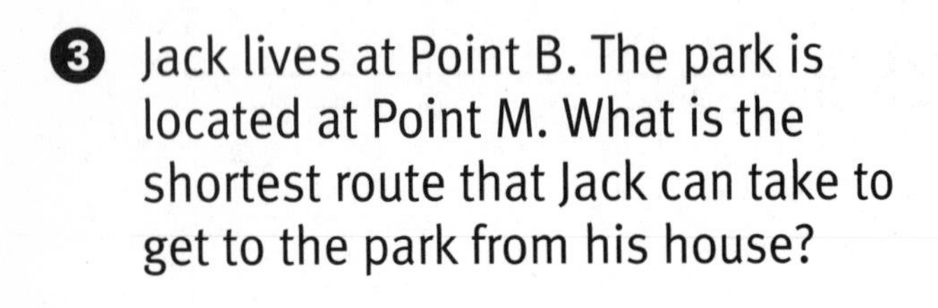

3. Jack lives at Point B. The park is located at Point M. What is the shortest route that Jack can take to get to the park from his house?

   ____

   ____

4. Jack leaves his house to go to the zoo. To get there, he goes 5 blocks east. Then he goes 1 block north. Finally, he goes 3 blocks east. What is the ordered pair that describes the location of the zoo? ________

   Which point describes this location? ________

To learn the answer, fill in the blanks by finding the letters in the graph that match the ordered pairs.

5. What kind of table has no legs?

___
(8, 6)

___ ___ ___ ___ ___ ___ ___ ___ ___ ___ ___ ___ ___ ___
(0, 7) (0, 0) (2, 10) (6, 2) (5, 1) (9, 9) (2, 10) (5, 1) (4, 5) (8, 6) (6, 2) (5, 1) (1, 4) (8, 3)

___ ___ ___ ___ ___
(6, 2) (8, 6) (2, 2) (2, 10) (10, 3)

LESSON 6.7

Name ____________________ Date __________

Functions and Graphing

The following function can be used to find out how many apples are needed to bake apple pies.

Complete the function table. Then complete the graph by plotting the rest of the ordered pairs from the table.

(x) Pies	(y) Apples
1	9
2	18
3	
4	
5	
6	
7	
8	
9	
10	

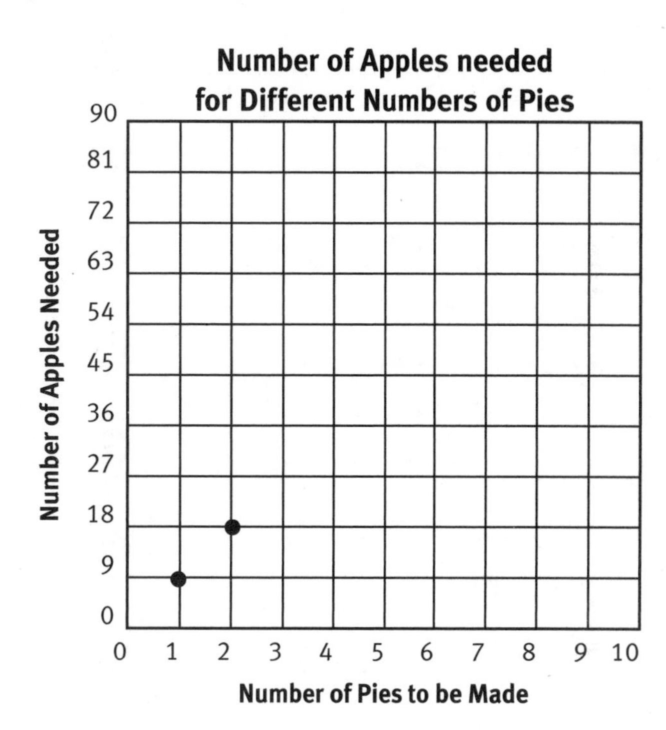

Answer each question using the table and graph above.

1. How many apples are needed to bake 3 pies? __________
2. How many pies can be made with 63 apples? __________
3. Using only the completed graph, can you find out how many apples are needed to make 9 pies? __________
4. Using only the completed graph, can you find out how many pies could be made if there were 54 apples? __________

LESSON 7.1

Name ______________________ Date ______________

Multiplying by 10, 100, and 1,000

Multiply mentally.

1. $10 \times 14 =$ ______
2. $26 \times 100 =$ ______
3. $10 \times 81 =$ ______
4. $1{,}000 \times 9 =$ ______
5. $74 \times 100 =$ ______
6. $3 \times 1{,}000 =$ ______
7. $10 \times 842 =$ ______
8. $81 \times 10 =$ ______
9. $4 \times 10 =$ ______
10. $1{,}000 \times 4 =$ ______
11. $20 \times 100 =$ ______
12. $10 \times 85 =$ ______
13. $7 \times 1{,}000 =$ ______
14. $1{,}000 \times 2 =$ ______
15. $6 \times 1{,}000 =$ ______
16. $100 \times 5 =$ ______
17. $10 \times 750 =$ ______
18. $10 \times 1{,}000 =$ ______
19. $82 \times 100 =$ ______
20. $50 \times 100 =$ ______

Solve. Show your work.

21. Patrick has 16 $10 bills. Does he have enough money to buy the sweater? ______

22. The coffee shop has 39 boxes of mugs. Each box contains 10 mugs. Are there enough mugs for 400 people? ______

LESSON 7.2

Name ______________________ Date ____________

Multiplying Two-Digit Numbers by One-Digit Numbers

Multiply. You may draw pictures to help.

1. $\begin{array}{r} 42 \\ \times 3 \\ \hline \end{array}$

2. $\begin{array}{r} 40 \\ \times 6 \\ \hline \end{array}$

3. $\begin{array}{r} 51 \\ \times 8 \\ \hline \end{array}$

4. $\begin{array}{r} 50 \\ \times 7 \\ \hline \end{array}$

5. $\begin{array}{r} 22 \\ \times 8 \\ \hline \end{array}$

6. $\begin{array}{r} 90 \\ \times 4 \\ \hline \end{array}$

7. $\begin{array}{r} 35 \\ \times 9 \\ \hline \end{array}$

8. $\begin{array}{r} 61 \\ \times 6 \\ \hline \end{array}$

9. $\begin{array}{r} 16 \\ \times 3 \\ \hline \end{array}$

10. $\begin{array}{r} 32 \\ \times 7 \\ \hline \end{array}$

11. $\begin{array}{r} 44 \\ \times 3 \\ \hline \end{array}$

12. $\begin{array}{r} 86 \\ \times 1 \\ \hline \end{array}$

Solve these problems.

13. There are 6 members in each row of a marching band. There are 32 rows. How many members are there in all? ____________

14. Ms. Carter ordered 8 boxes of colored pencils for her class. Each box had 35 pencils. How many pencils did Ms. Carter order? ____________

15. Karen sold 47 ice-cream cones for \$2 each. How much money did she get from selling ice-cream cones? ____________

LESSON 7.3

Name ______________________ Date __________

Multiply Three-Digit Numbers by One-Digit Numbers

Multiply. You may draw pictures to help.

1. 234×3
2. 786×0
3. 705×7
4. 444×6
5. 928×2
6. 245×6
7. 205×9
8. 213×7
9. 438×2
10. 701×6
11. 125×8
12. 260×4

Solve these problems.

13. In one week, Ben, Joseph, and Andy each collected 125 stamps. How many stamps did they collect in all? __________

14. Willard jogged 7 times around a 250 meter race track. How many meters did he jog in all? __________

15. Franklin had 472 buttons in a drawer. Then he put in 9 more buttons. How many buttons does he have now? __________

LESSON 7.6

Name ________________________ Date ____________

Reasonable Answers to Division Problems

Is each of the following possible? If yes, write the answer in a sentence. If no, explain why not.

1. There are 20 students in Mrs. Baker's class. She wants to put her students into 4 equal work groups. How many students will be in each group?

   ____

2. Paulo gave 55 marbles to 4 of his friends. He gave away all of his marbles, and each friend had an equal number of marbles. How many marbles did each friend get?

   ____

3. Alice was selling boxes of chocolates to raise money for her school. She sold an equal number of boxes to 8 people. Altogether, she sold 26 boxes. How many boxes did each person buy?

   ____

4. Mr. Brodsky is taking his class on a field trip. There are 23 students in his class. They will ride in cars that can hold 5 students each. How many cars will be needed?

   ____

In each problem, two of the answers are clearly incorrect, and one is correct. Ring the correct answer.

5. $6\overline{)360}$

   a. 600 **b.** 60 **c.** 6

6. $10\overline{)115}$

   a. 11 R5 **b.** 1 R15 **c.** 101 R5

7. $10\overline{)1000}$

   a. 100 **b.** 10 **c.** 1

8. $7\overline{)250}$

   a. 3 R10 **b.** 350 R5 **c.** 35 R5

9. $10\overline{)771}$

   a. 77 R10 **b.** 77 R1 **c.** 7 R1

10. $8\overline{)128}$

   a. 6 **b.** 16 **c.** 160

LESSON 7.7

Name ____________________ Date __________

Dividing Two-Digit Numbers by One-Digit Numbers

Divide. Keep records to show your answer.

1. $3\overline{)39}$
2. $7\overline{)98}$
3. $6\overline{)78}$
4. $4\overline{)60}$
5. $4\overline{)72}$
6. $5\overline{)75}$
7. $6\overline{)84}$
8. $6\overline{)90}$
9. $2\overline{)42}$
10. $2\overline{)78}$
11. $3\overline{)84}$
12. $8\overline{)96}$
13. $5\overline{)65}$
14. $3\overline{)93}$
15. $4\overline{)96}$
16. $2\overline{)88}$

Solve these problems. Be sure to label your answer correctly.

17. Bill set 5 equal weights on a scale. The scale showed their total weight was 70 pounds. How many pounds was each weight? __________
18. Zeke counted 58 socks in his drawer. How many pairs of socks did he have? __________
19. Dinner for 3 people costs $48. If they split the cost evenly between them, how much does each person pay? __________

LESSON 7.8

Name ______________________ Date __________

Dividing Three-Digit Numbers by One-Digit Numbers

Divide. Keep records to show your answer.

1. $4\overline{)860}$
2. $3\overline{)678}$
3. $5\overline{)945}$
4. $5\overline{)185}$
5. $9\overline{)459}$
6. $7\overline{)406}$
7. $5\overline{)480}$
8. $3\overline{)837}$
9. $2\overline{)648}$
10. $2\overline{)992}$
11. $8\overline{)360}$
12. $9\overline{)711}$
13. $4\overline{)228}$
14. $7\overline{)644}$
15. $8\overline{)512}$
16. $9\overline{)828}$
17. $6\overline{)678}$
18. $6\overline{)864}$
19. $7\overline{)497}$
20. $4\overline{)260}$

LESSON 7.9

Name ______________________________ Date ____________

Problem-Solving Applications

Solve these problems. If it is not possible to solve the problem, tell why.

1. The biking club has 6 teams with 4 members on each team. How many members are there altogether? ____________

2. A bag of popcorn costs 8¢. Terry has 35¢. How many bags of popcorn can she buy? ____________

3. Jules bought a box of pencils and a notebook. Altogether, he spent $5. How much did the notebook cost? ____________

4. Bananas cost 9¢ each. How much will 6 bananas cost? ____________

5. Jim was 46 inches tall. He grew 6 inches this year. How tall is Jim now? ____________

6. A school basketball club has 36 players. The coach wants to make 4 equal teams. How many players will be on each team? ____________

7. Polly has $28. She buys a gift for her mother. Now she has $6. How much did she spend on her mother's gift? ____________

8. Gilberto gets 10¢ for each chore he does in a day. Gilberto earned 80¢ in 1 day. How many chores did Gilberto do? ____________

9. Mr. Paul rode his bike 13 kilometers and then back home. How far did Mr. Paul ride? ____________

LESSON 8.1

Name ______________________ Date __________

Fractions of Geometric Figures

What fraction of each figure is shaded?

1. 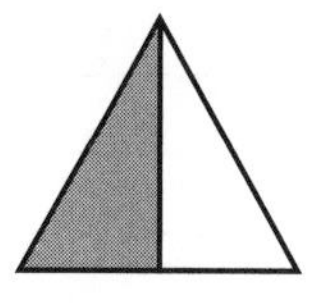

______

2. 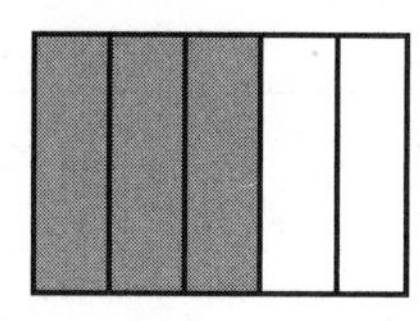

______

3. 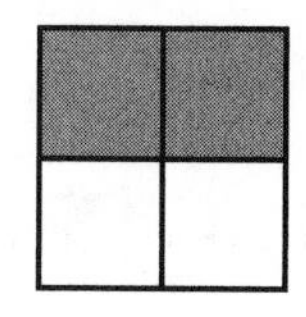

______

4. 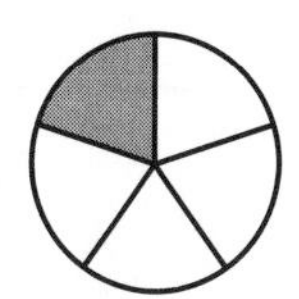

______

5. 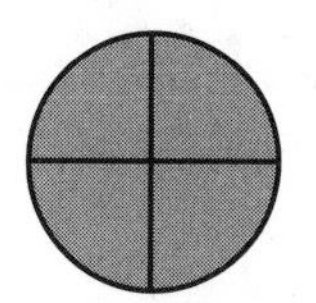

______

6. 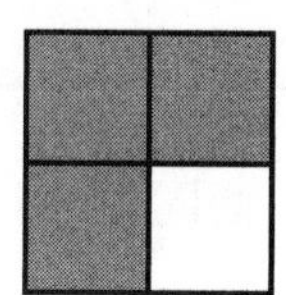

______

Shade each figure for the given fraction.

7. Shade $\frac{3}{4}$ of the triangle.

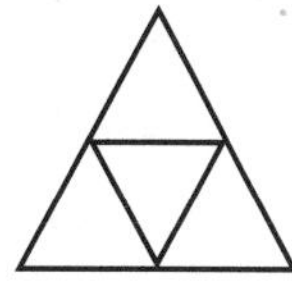

8. Shade $\frac{4}{5}$ of the circle.

9. Shade $\frac{2}{5}$ of the rectangle.

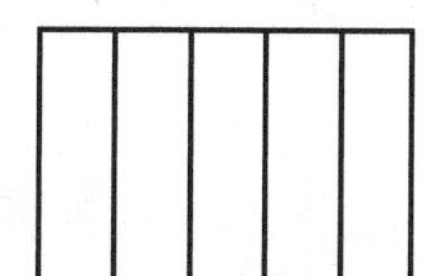

10. Copy or trace this figure 4 times. Then shade $\frac{2}{4}$ in five different ways.

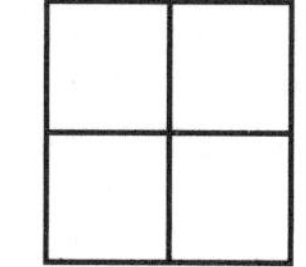

LESSON 8.2

Name ______________________ Date ____________

Fractions of Linear Measure

What fraction of the line is shaded?

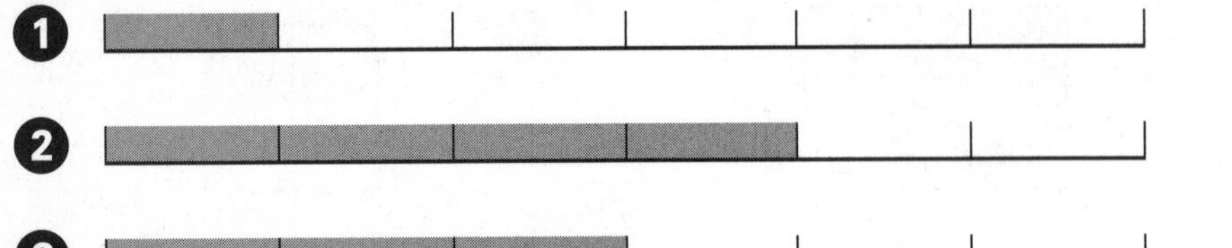

1. ________
2. ________
3. ________
4. ________
5. ________

Each ruler below is divided into twelve equal sections. Each section is $\frac{1}{12}$ of a foot. What fraction of a foot is each ribbon?

6. ________

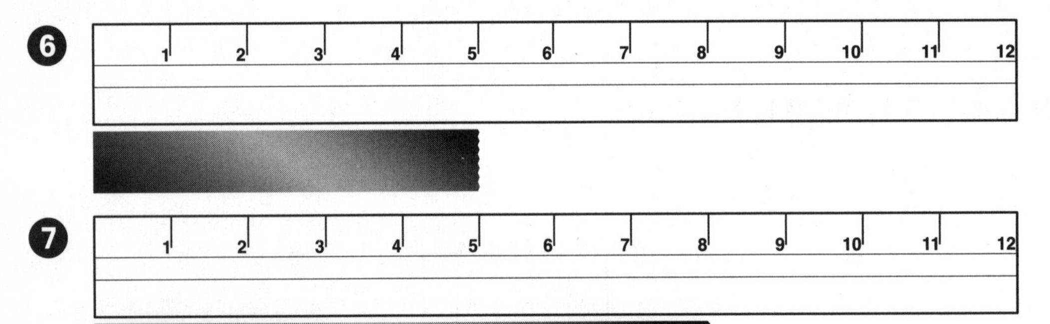

7. ________

8. ________

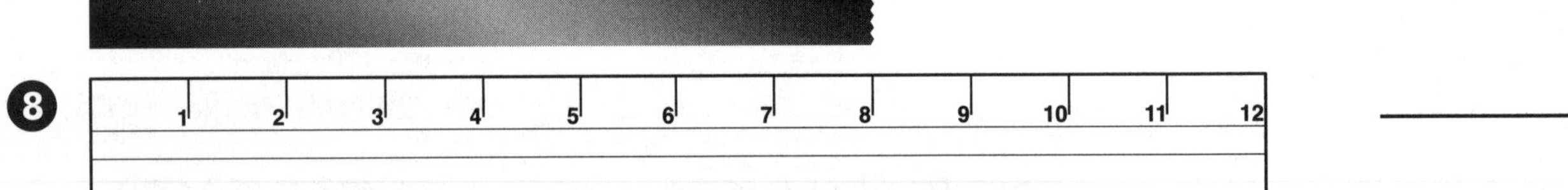

9. ________

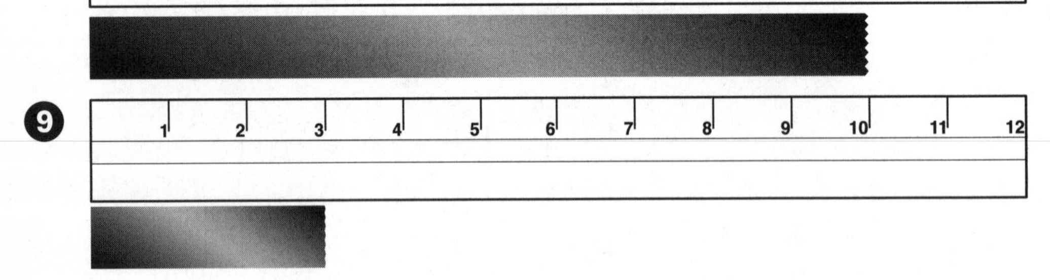

Solve. Show your work.

10. Miguel's frog jumped 7 feet. The frog race was 12 feet from start to finish. What fraction of the race did Miguel's frog jump? ________

LESSON 8.3

Name ______________________ Date ____________

Fractions of Sets

Use the picture to answer the questions.

1. What fraction of the fruit is bananas? ________
2. What fraction of the fruit is apples? ________
3. What fraction of the fruit includes all the cherries and all the apples? ________
4. What fraction of the fruit includes all the bananas and all the apples? ________
5. What fraction of the fruit is not apples? ________

You may use manipulatives to help you solve these problems.

6. 

How many cars are in each group? ________

$\frac{1}{3}$ of 15 is ________

$\frac{1}{5}$ of 15 is ________

7. $\frac{2}{5}$ of 10 is ________
8. $\frac{3}{5}$ of 10 is ________
9. $\frac{1}{3}$ of 21 is ________
10. $\frac{2}{5}$ of 30 is ________

LESSON 8.4

Name ______________________ Date ____________

Fractions of Time

How many minutes?

1. 1 hour = ____________ minutes
2. $\frac{3}{6}$ of an hour = ____________ minutes
3. $\frac{1}{3}$ of an hour = ____________ minutes
4. $\frac{2}{6}$ of an hour = ____________ minutes
5. $\frac{3}{4}$ of an hour = ____________ minutes
6. $\frac{2}{4}$ of an hour = ____________ minutes
7. $\frac{2}{3}$ of an hour = ____________ minutes
8. $\frac{4}{4}$ of an hour = ____________ minutes
9. $\frac{1}{6}$ of an hour = ____________ minutes
10. $\frac{1}{4}$ of an hour = ____________ minutes

Which is longer?

11. $\frac{1}{3}$ of an hour or $\frac{2}{4}$ of an hour? ____________
12. $\frac{1}{2}$ of an hour or $\frac{1}{3}$ of an hour? ____________
13. $\frac{3}{3}$ of an hour or $\frac{4}{4}$ of an hour? ____________
14. $\frac{3}{4}$ of an hour or $\frac{2}{3}$ of an hour? ____________
15. $\frac{1}{4}$ of an hour or $\frac{1}{3}$ of an hour? ____________
16. $\frac{2}{2}$ of an hour or $\frac{1}{2}$ of an hour? ____________

LESSON 8.5

Name ______________________ Date ____________

Equivalent Fractions

What fraction of each circle is shaded?

1. 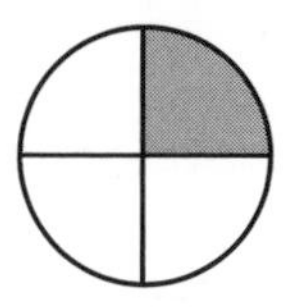
______

2. 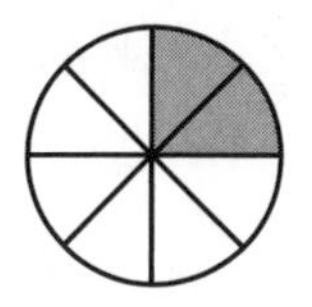
______

3. 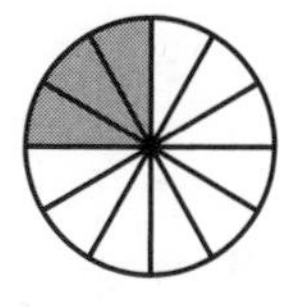
______

4. 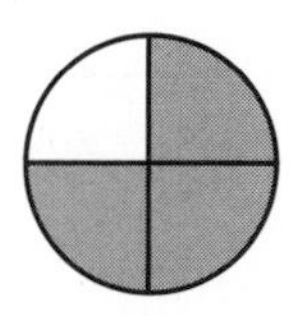
______

5. 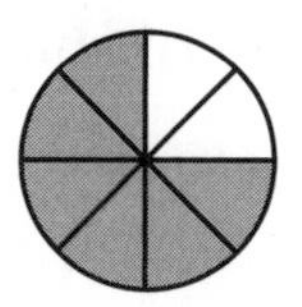
______

6. 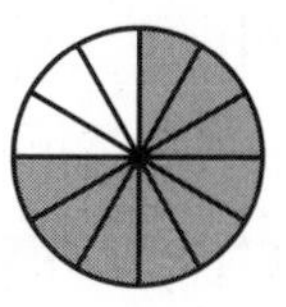
______

What fraction of each rectangle is shaded?

7. 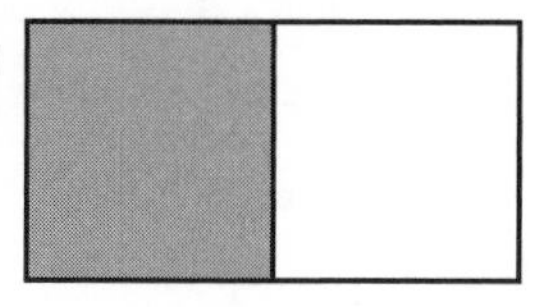
______

8. 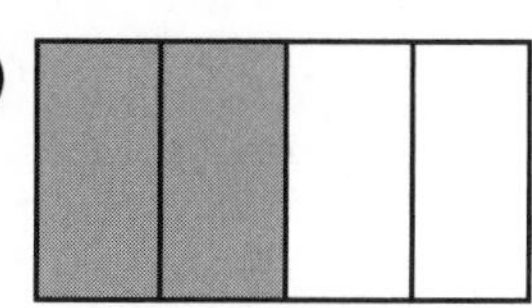
______

9. 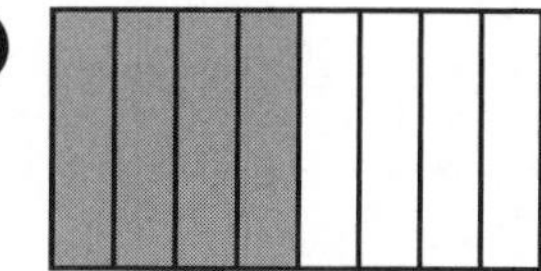
______

10. 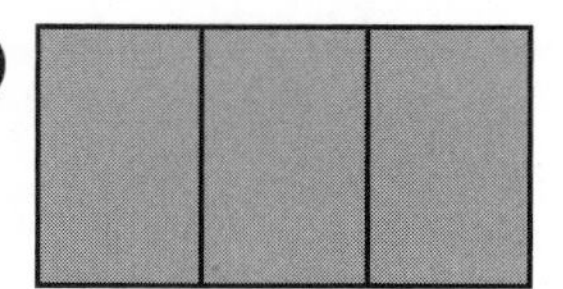
______

11. 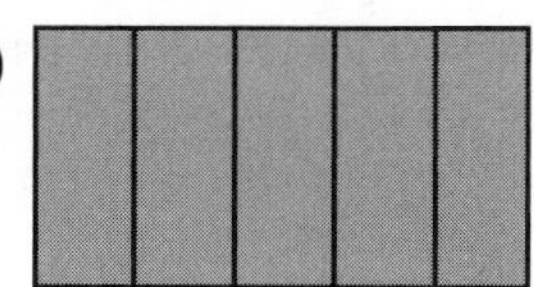
______

12. 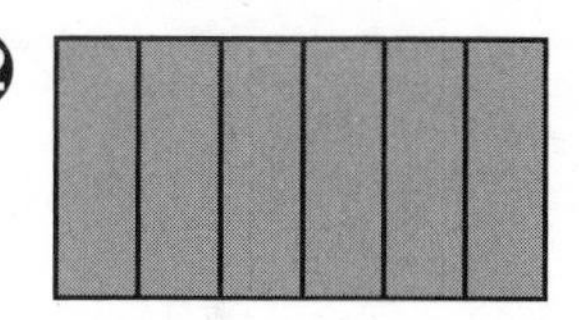
______

Solve.

13. List or draw 2 fractions that represent $\frac{2}{3}$.

____

LESSON 8.6

Name ____________________ Date __________

Comparing Fractions

Answer these questions.

1. Which is bigger, $\frac{1}{4}$ of the carrot or $\frac{1}{5}$ of the carrot? __________
2. Which is bigger, $\frac{2}{4}$ of the peach or $\frac{2}{3}$ of the peach? __________
3. Which is bigger, $\frac{1}{2}$ of the cake or $\frac{4}{5}$ of the cake? __________
4. Which is bigger, $\frac{3}{4}$ of the orange or $\frac{2}{2}$ of the orange? __________
5. Which is bigger, $\frac{3}{5}$ of the sandwich or $\frac{1}{3}$ of the sandwich? __________
6. Which is bigger, $\frac{1}{3}$ of the strawberry or $\frac{2}{6}$ of the strawberry? __________

Write $<$, $>$, or $=$ to make a true statement.

7. $\frac{1}{3}$ ______ $\frac{1}{5}$
8. $\frac{2}{4}$ ______ $\frac{4}{5}$
9. $\frac{3}{4}$ ______ $\frac{2}{5}$
10. $\frac{2}{2}$ ______ $\frac{4}{4}$
11. $\frac{3}{4}$ ______ $\frac{3}{5}$
12. $\frac{2}{3}$ ______ $\frac{2}{4}$
13. $\frac{2}{5}$ ______ $\frac{1}{2}$
14. $\frac{3}{5}$ ______ $\frac{3}{3}$
15. $\frac{5}{5}$ ______ $\frac{2}{2}$
16. $\frac{1}{2}$ ______ $\frac{3}{5}$

17. Which is bigger, $\frac{3}{4}$ or $\frac{2}{3}$? Explain your answer.

____

____

LESSON 8.7

Name ______________________ Date ____________

Adding and Subtracting Fractions

Solve these problems.

1. A recipe uses $\frac{1}{3}$ cup of flour. Scott wants to double the recipe. How much flour will he use? ____________

2. Mira collects sports cards. She knows that $\frac{2}{5}$ of her collection is baseball cards, and $\frac{1}{5}$ is football cards. What fraction of her collection is baseball cards and football cards altogether? ____________

Write <, >, or = to make a true statement.
The pictures may help you.

3. 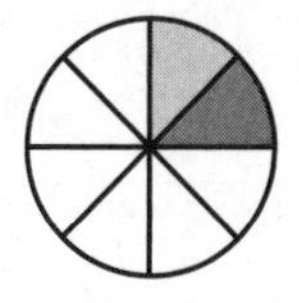

$\frac{1}{8} + \frac{1}{8}$ ________ $\frac{1}{2}$

4. 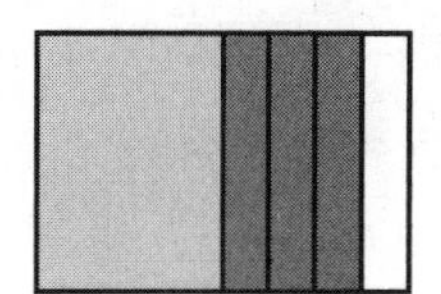

$\frac{1}{2} + \frac{3}{8}$ ________ $\frac{4}{4}$

5. 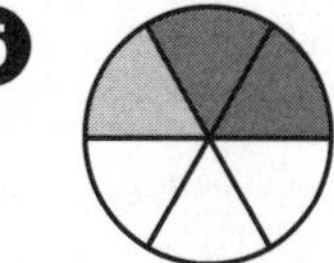

$\frac{1}{6} + \frac{2}{6}$ ________ $\frac{1}{2}$

6. 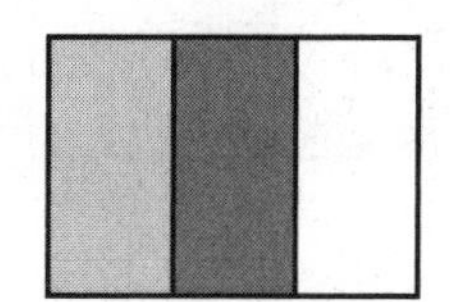

$\frac{1}{3} + \frac{1}{3}$ ________ $\frac{3}{4}$

Solve.

7. $\frac{4}{5} - \frac{1}{5} =$ ________

8. $\frac{2}{7} + \frac{4}{7} =$ ________

9. $\frac{2}{3} - \frac{3}{6} =$ ________

10. $\frac{3}{6} - \frac{1}{2} =$ ________

11. $\frac{4}{6} + \frac{2}{6} =$ ________

12. $\frac{2}{4} + \frac{1}{4} =$ ________

13. $\frac{1}{2} - \frac{1}{4} =$ ________

14. $\frac{1}{7} + \frac{3}{7} =$ ________

LESSON 8.8

Name ______________________ Date ____________

Fractions Greater Than a Whole

Name the fraction represented by the shaded sections in two ways.

1. 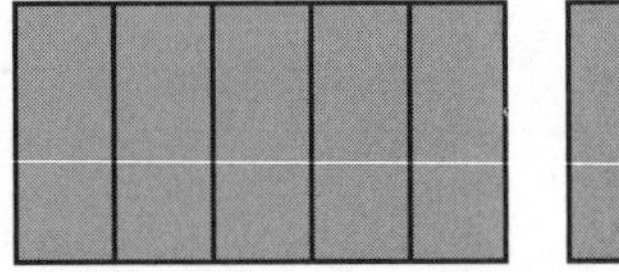 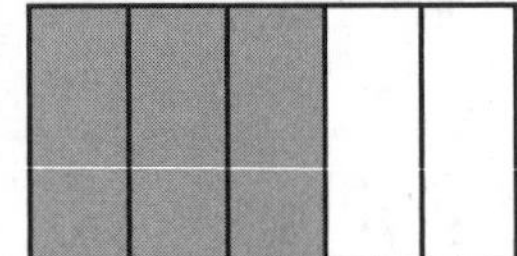

____

2. 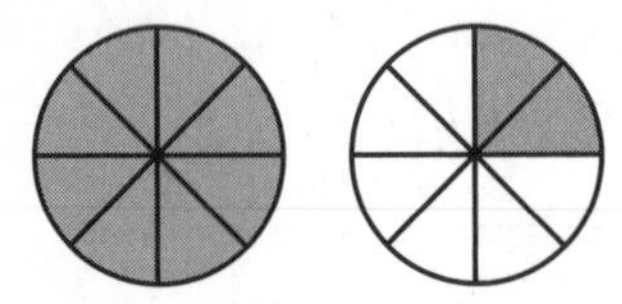

____

3. 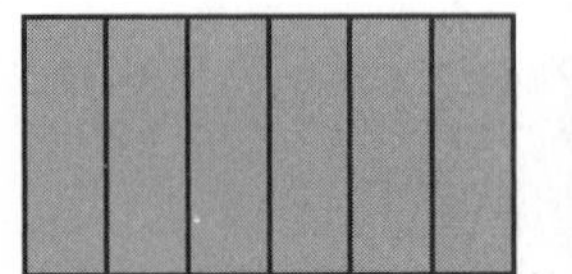 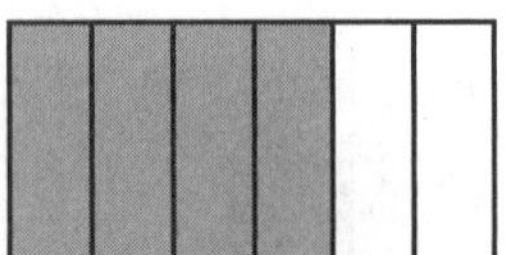

____

4. 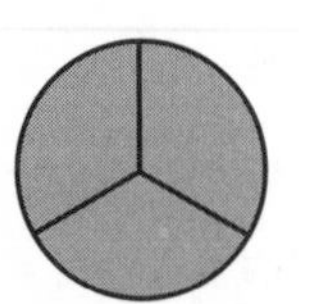 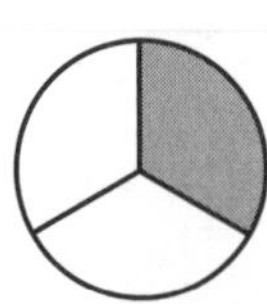

____

5. 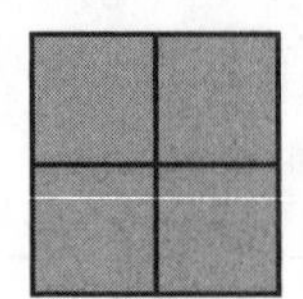 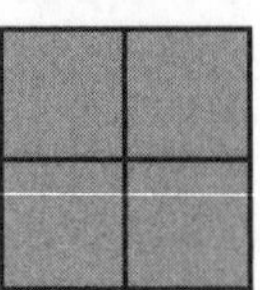

____

6. 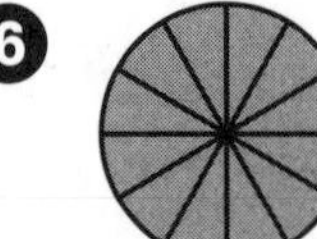 

____

7. 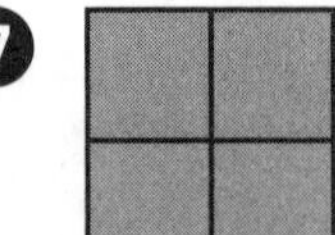 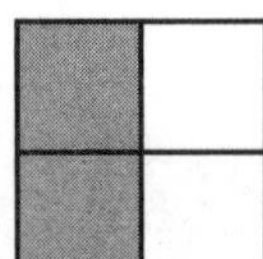

____

8. 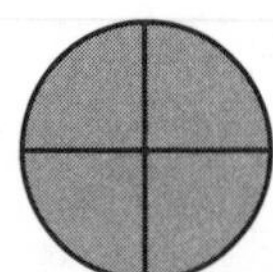 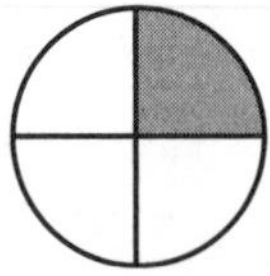

____

LESSON 8.9

Name ______________________ Date __________

Comparing Fractions Greater Than a Whole

Write <, >, or = to make a true statement. Use fraction circles, fraction tiles, or drawings if needed.

1. $2\frac{1}{3}$ ______ $1\frac{3}{4}$
2. $3\frac{2}{4}$ ______ $3\frac{1}{2}$
3. $5\frac{3}{4}$ ______ $6\frac{1}{4}$
4. $2\frac{1}{5}$ ______ $1\frac{4}{5}$
5. $7\frac{1}{3}$ ______ $6\frac{3}{4}$
6. $6\frac{2}{4}$ ______ $7\frac{1}{4}$
7. $4\frac{2}{5}$ ______ $3\frac{4}{5}$
8. $6\frac{2}{3}$ ______ $6\frac{2}{4}$
9. $4\frac{2}{3}$ ______ $4\frac{3}{4}$
10. $\frac{4}{5}$ ______ $1\frac{1}{5}$
11. $3\frac{1}{2}$ ______ $3\frac{2}{3}$
12. $5\frac{1}{2}$ ______ $5\frac{2}{4}$

Solve. **Show your work.**

13. Jordan used $4\frac{3}{4}$ cups of flour in his recipe. Belinda used $4\frac{2}{3}$ cups of flour in her recipe. Who used more flour? Explain how you found your answer.

____

____

LESSON 8.10

Name ______________________ Date ____________

Tenths and Hundredths

Use the circle below to help answer the questions. It is divided into 100 equal parts.

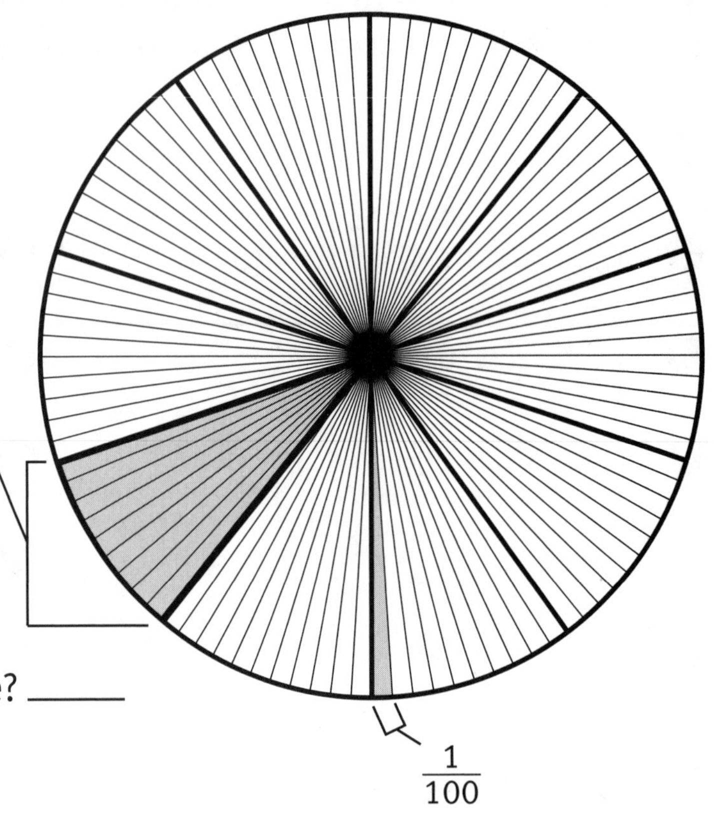

1. In $\frac{4}{10}$ of the circle, how many $\frac{1}{100}$ sections are there? ______
2. In $\frac{7}{10}$ of the circle, how many $\frac{1}{100}$ sections are there? ______
3. In $\frac{30}{100}$ sections of the circle, how many tenths of the circle are there? ______
4. In $\frac{90}{100}$ sections of the circle, how many tenths of the circle are there? ______
5. What is another fraction to represent $\frac{8}{10}$ of the circle? __________

Solve.

6. What is $\frac{1}{10}$ of 400? ______
7. What is $\frac{2}{10}$ of 400? ______
8. What is $\frac{2}{10}$ of 500? ______
9. What is $\frac{3}{10}$ of 200? ______
10. What is $\frac{3}{10}$ of 300? ______
11. What is $\frac{4}{10}$ of 100? ______
12. What is $\frac{4}{10}$ of 200? ______
13. What is $\frac{5}{10}$ of 300? ______
14. What is $\frac{10}{10}$ of 400? ______
15. What is $\frac{9}{10}$ of 100? ______

LESSON 8.11

Name ______________________ Date ____________

Percents and Hundredths

Write the fraction for the following percents. Remember, a percent is a fraction with a denominator of 100.

1. 18% ________
2. 39% ________
3. 88% ________
4. 7% ________

Write the percent for the following fractions.

5. $\frac{10}{100}$ ________
6. $\frac{99}{100}$ ________
7. $\frac{1}{100}$ ________
8. $\frac{67}{100}$ ________

Use the graph to answer the following questions. Remember, a percent is a fraction with a denominator of 100.

9. What fraction is each section of this circle graph?

   a. Soccer ________

   b. Swimming ________

   c. Basketball ________

   d. Baseball ________

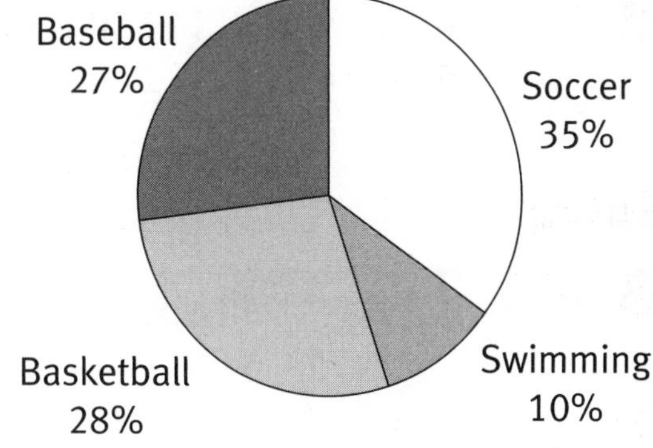

10. What was the most popular sport at Atholton Elementary School? ________

11. Of the following percents, 0%, 25%, 50%, or 100%, which would you choose to describe about how many students voted for baseball? Explain.

____

____

LESSON 9.1

Name ____________________ Date __________

Metric Length

Metric Units of Length

There are 1,000 millimeters in 1 meter. 1,000 mm = 1 m
There are 100 centimeters in 1 meter. 100 cm = 1 m
There are 1,000 meters in 1 kilometer. 1,000 m = 1 km

Convert each measure. Write the new measure.

1. 300 cm = __________ m
2. 4,000 mm = __________ m
3. 5 m = __________ cm
4. 20 mm = __________ cm
5. 500 mm = __________ cm
6. 9,000 mm = __________ m

Write the unit that is most reasonable.
Use *millimeters, centimeters, meters,* or *kilometers.*

7. Mark's father is 183 ____________________ tall.
8. The map says that the distance from Biloxi to Jackson is about 210 ____________________ .
9. A ladybug is about 8 ____________________ long.
10. Ken hit a baseball about 60 ____________________ .

Solve each problem. Show your work.

11. Armando ran 4 laps around a 750-meter racetrack at his school. How many kilometers did he run? __________
12. A goliath beetle can grow up to 150 millimeters in length. How long can a goliath beetle grow in centimeters? __________

LESSON 9.2

Name ______________________ Date __________

Customary Length

Customary Units of Length

There are 12 inches in 1 foot.	12 in. = 1 ft
There are 3 feet in 1 yard.	3 ft = 1 yd
There are 36 inches in 1 yard.	36 in. = 1 yd

Convert each measure. Write the equivalent measure.

1. 2 yards = __________ inches
2. 12 feet = __________ yards
3. 30 yards = __________ feet
4. 6 feet = __________ inches
5. 24 inches = __________ feet
6. 60 feet = __________ yards

Solve each problem. Show your work.

7. Wilson's dining room table is 6 feet long. Convert in yards. __________
8. Dorinda's jump rope is 3 feet long. Convert in inches. __________
9. Isaac's backyard is 10 yards deep. Convert in feet. __________
10. The height of the flagpole outside Samuel's school is 24 feet. What is the height of the flagpole in yards? __________
11. In a football game, Gerald ran 12 yards before being tackled. How many feet did he run? __________
12. After the game, Gerald jogged a victory lap around the edge of the football field. The field was 120 yards long by 53 yards wide.

    a. How many yards did Gerald jog? __________

    b. How many feet did Gerald jog? __________

LESSON 9.3

Name ______________________ Date ____________

Metric Weight

Metric Units of Weight

There are 1,000 grams in 1 kilogram. 1,000 g = 1 kg

Convert each measure. Write the equivalent measure.

1. ____________ kg = 4,000 g
2. 7 kg = ____________ g
3. 12 kg = ____________ g
4. $1\frac{1}{2}$ kg = ____________ g

Ring the answer that makes the most sense.

5.  weighs about ____________ grams.

   a. 2 **b.** 20 **c.** 200

6. 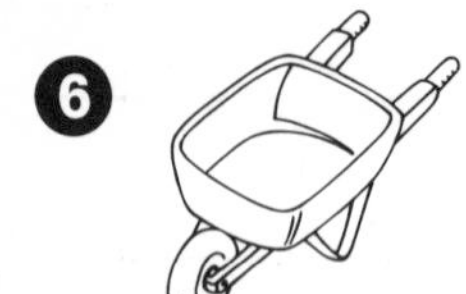 weighs about ____________ kilograms.

   a. 15 **b.** 150 **c.** 1,500

7.  weighs about ____________ grams.

   a. 3 **b.** 350 **c.** 3,500

Solve each problem.

8. Melanie caught 4 bullfrogs in a pond. Each one weighs about 750 grams. Altogether, about how many kilograms do Melanie's bullfrogs weigh? ____________

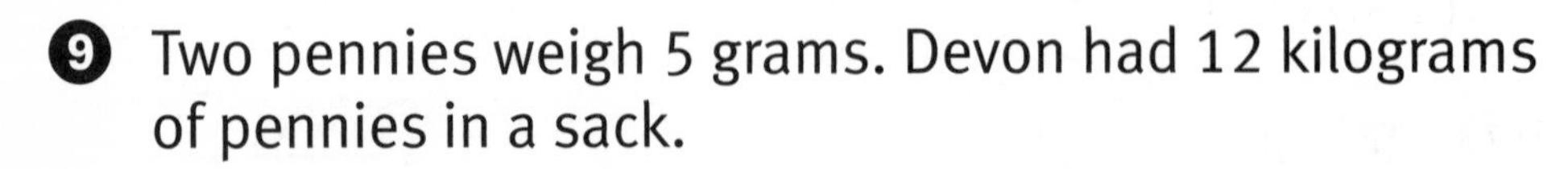

9. Two pennies weigh 5 grams. Devon had 12 kilograms of pennies in a sack.

   a. How many pennies were in the sack? ____________

   b. How much were the pennies worth in dollars? ____________

LESSON 9.4

Name ______________________ Date __________

Customary Weight

Customary Units of Weight

There are 16 ounces in 1 pound. 16 oz = 1 lb

Convert each measure. Write the equivalent measure.

1. 48 oz = __________ lbs
2. 5 lbs = __________ oz
3. 16 lbs = __________ oz
4. 248 oz = __________ lbs
5. 1,840 oz = __________ lbs
6. 7 lbs = __________ oz

Ring the answer that makes the most sense.

7.  weighs about __________ pounds.

   a. 8 **b.** 80 **c.** 800

8.  weighs about __________ ounces.

   a. 500 **b.** 55 **c.** 5

9.  weighs about __________ ounces.

   a. 160 **b.** 16 **c.** 1

10.  weighs about __________ pounds.

    a. 1 **b.** 11 **c.** 150

LESSON 9.5

Name ______________________ Date ____________

Metric Capacity

Metric Units of Capacity

There are 1,000 milliliters in 1 liter. 1,000 mL = 1 L

Convert each measure. Write the equivalent measure.

1. 3 L = ____________ mL
2. 6 L = ____________ mL
3. $3\frac{1}{2}$ L = ____________ mL
4. ____________ L = 7,000 mL
5. ____________ L = 500 mL
6. ____________ L = 10,000 mL

Write the unit that makes sense. Use *milliliters* or *liters*.

7. 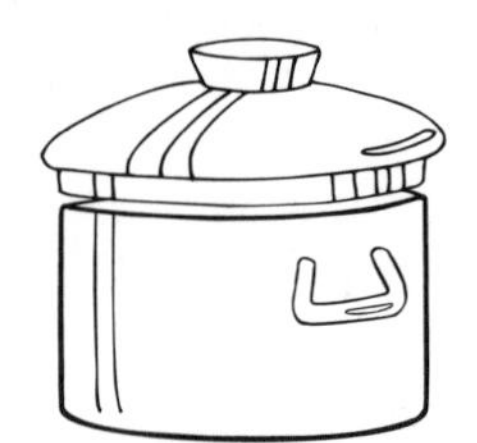

about 8 ____________ of sauce

10. 

about 240 ____________ of water

8. 

about 50 ____________ of ink

11. 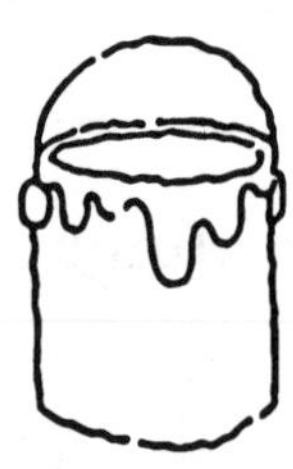

about 5 ____________ of paint

9. 

about 150 ____________ of tea

12. 

about 15 ____________ of soup

LESSON 9.6

Name ______________________ Date ____________

Customary Capacity

Customary Units of Capacity

There are 2 cups in 1 pint.
There are 2 pints in 1 quart.
There are 4 quarts in 1 gallon.

Convert **each measure. Write the new measure.**

1. 2 pints = ____________ cups
2. $\frac{1}{2}$ gallon = ____________ quarts
3. ____________ cups = 1 gallon
4. ____________ pints = 5 quarts
5. ____________ quarts = 8 cups
6. 6 pints = ____________ quarts
7. $\frac{1}{4}$ gallon = ____________ pints
8. 8 cups = ____________ pints

Draw **a line from each measure of capacity to a container on the table that can hold that much.**

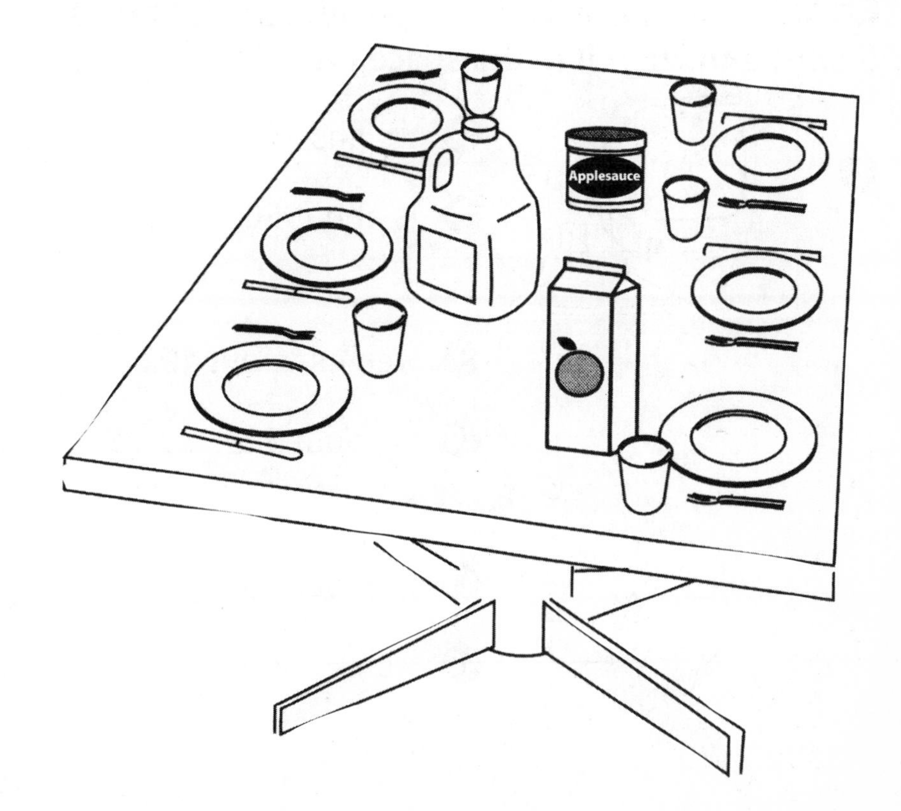

9. About 1 gallon
10. About 1 quart
11. About 1 cup

12. About 1 pint

LESSON 9.7

Name ______________________ Date ____________

Choosing the Correct Unit

Choose the correct unit of measurement, ***inches, feet, yards, ounces,*** or ***pounds,*** and write it in the space provided.

1. weighs about 5 ____________
2. is about 8 ____________ wide

3. is about 2 ____________ wide
4. is about 4 ____________ high

5. is about 15 ____________ long
6. weighs about 25 ____________

Choose the correct unit of measurement, ***meters, centimeters, kilograms,*** or ***grams,*** and write it in the space provided.

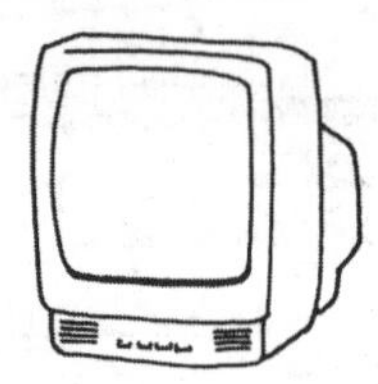

7. weighs about 40 ____________
8. is about 50 ____________ wide

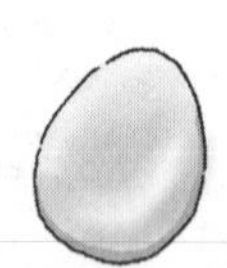

9. weighs about 40 ____________
10. is about 5 ____________ long

11. is about 30 ____________ tall
12. weighs about 5 ____________

LESSON 9.8

Name ______________________ Date __________

Measuring Elapsed Time

Train Schedule from Baltimore, MD, to New York, NY		
Train	Departs	Arrives
A	6:20 A.M.	8:45 A.M.
B	10:40 A.M.	1:05 P.M.
C	1:20 P.M.	4:15 P.M.
D	5:10 P.M.	9:00 P.M.
E	5:45 P.M.	8:55 P.M.
F	10:45 P.M.	12:50 P.M.

Use the chart to answer the following questions.

1. Which train has the longest travel time? __________
2. Which train has the shortest travel time? __________
3. Alex needs to travel from Baltimore to New York, but he does not want to be on the train for more than 3 hours. Which trains could he take? __________
4. Train B left the station 20 minutes late. What time did the train leave? __________
5. Train D arrived 15 minutes early. What time did the train arrive? __________
6. Mr. Guzman needs to be in New York for a business meeting by 5:00 P.M., but he cannot leave Baltimore until after 9:00 A.M. Which trains can he take? __________
7. Train C departed and arrived on time. How many hours was the trip? __________
8. Immanuel took Train A from Baltimore to New York. It left 5 minutes early, but arrived 15 minutes late.
   a. What time was it when he left? __________
   b. What time was it when he arrived? __________
   c. How many hours did he spend traveling? __________

LESSON 9.9

Name ______________________ Date ____________

Understanding the Metric System

Length Units	Weight Units	Capacity Units	U.S. Currency
millimeter (mm)	milligram (mg)	milliliter (mL)	mill
centimeter (cm)	centigram (cg)	centiliter (cL)	cent
decimeter (dm)	decigram (dg)	deciliter (dL)	dime
meter (m)	gram (g)	liter (L)	dollar bill ($1)
decameter (dam)	dekagram (dag)	dekaliter (daL)	10-dollar bill
hectometer (hm)	hectogram (hg)	hectoliter (hL)	100-dollar bill
kilometer (km)	kilogram (kg)	kiloliter (kL)	1,000-dollar bill

Use the table to answer each question.

1. How many milliliters are in 1 liter? ____________
2. How many 10-dollar bills are in $100? ____________
3. How many dimes are in $100? ____________
4. How many centigrams are in 1 gram? ____________
5. How many meters are in 1 kilometer? ____________
6. How many decigrams are in 1 gram? ____________
7. How many dimes are in $1? ____________
8. How many cents are in $10? ____________
9. How many deciliters are in 1 kiloliter? ____________

LESSON 10.1

Name ______________________ Date ____________

Where Do We See Decimals?

Solve these problems. Show your work.

1. A rectangular lunch table is 203 centimeters long and 115 centimeters wide. What is the perimeter of the lunch table? Write your answer in meters. ____________

2. One bag of potatoes costs 197 cents. How much do 4 bags cost? Write your answer using dollars. ____________

3. The distance around Nature Park is 2.8 kilometers. On Tuesday, Travis jogged around the park 3 times. How many kilometers did Travis jog? ____________

4. Riverside Elementary uses 435 grams of peanut butter each day on lunches and snacks. How much peanut butter do they use in 5 days? Write your answer in kilograms. ____________

5. Claudia wants to buy tulips for her garden. The flower shop has a dozen tulips on sale for $5.25. Does Claudia have enough to buy 2 dozen tulips? ____________

6. Mr. Barone needs 1.3 kilograms of tomato sauce to make a pasta dish. He plans to buy the sauce shown on the advertisement below. How many cans of sauce does Mr. Barone need to buy? ____________

LESSON 10.2

Name ______________________ Date __________

Decimals and Money

Write the equivalent amounts.

1. 6 dimes = ______ ¢
2. 13 dimes = $______
3. ______ dimes = $40
4. $______ = 30 dimes
5. 600 dimes = ______ ¢
6. $______ = 8 dimes

Solve. Make the least number possible in each case.

7. $______ = $3, 11 dimes, and 4¢
8. $29.75 = $______, ______ dimes, and ______ ¢
9. $34.21 = $______, ______ dimes, and ______ ¢

Solve these problems. Show your work.

10. If you take 437 dimes to the bank to exchange for one-dollar bills, how many one-dollar bills will you get? Will you have any dimes left over?

____

11. Carlos buys and sells bottles of water during the summer. When the temperature is below 90°F, he sells water bottles for $1.25 each. When the temperature is above 90°F, he sells them for $1.50 each.

   a. Carlos bought a case of 48 water bottles for $40. He sold all the water bottles on a day when the temperature was 84°F. How much profit did he make that day? ______________________________

   b. Carlos bought another case of 48 water bottles for $40, and sold them all on a day when the temperature was 96°F. How much profit did he make then?

   ____

   c. Is it fair to sell the bottles at a higher price on hotter days? Why or why not?

   ____

LESSON 10.3

Name ______________________ Date __________

Decimals and Metric Measures

There are 10 decimeters in one meter. 10 dm = 1 m

There are 10 centimeters in one decimeter. 10 cm = 1 dm

Write the equivalent measures.

1. 40 dm = ______ m
2. 70 dm = ______ m
3. 9 m = ______ dm
4. 90 m = ______ dm
5. 37 dm = ______ m and ______ dm
6. 900 dm = ______ m
7. 10.3 m = ______ m and ______ dm
8. 9.7 m = ______ dm
9. 8 m = ______ dm
10. 250 dm = ______ m
11. 80 dm = ______ m
12. 12 m = ______ dm
13. 230 dm = ______ m
14. 48 dm = ______ m and ______ dm
15. 22.3 m = ______ dm
16. 21 dm = ______ m and ______ dm

Solve these problems. Show your work.

Bobby is making a poster for his room. He has a large piece of poster board in the shape of a rectangle that is 80 centimeters wide and 140 centimeters long.

17. What is the perimeter of the poster board? Write your answer in centimeters and decimeters. ______

18. Bobby has 4.1 meters of colored ribbon. Does he have enough ribbon to decorate the entire perimeter of his poster? ______

19. A frame at the craft store measures 0.9 meters by 1.5 meters. Is this frame large enough to hold Bobby's poster? ______

LESSON 10.4

Name ______________________ Date ____________

Comparing Decimals and Fractions

Write these fractions as decimals.

1. $\frac{4}{10} =$ ________
2. $\frac{83}{10} =$ ________
3. $\frac{37}{100} =$ ________
4. $\frac{95}{100} =$ ________
5. $\frac{6}{10} =$ ________
6. $2\frac{9}{100} =$ ________

Write these decimals as fractions.

7. 7.2 = ________
8. 0.13 = ________
9. 1.46 = ________
10. 0.08 = ________
11. 5.9 = ________
12. 0.40 = ________

Solve these problems. Show your work.

13. Salena wants to buy a new bike that costs $218.00. She makes $6.40 an hour working at the toy store.

   a. This week, Salena plans to work $5\frac{1}{4}$ hours on Monday, Wednesday, and Friday. How much money will she earn this week? Is that enough to buy the bike?

   ____

   b. If Salena keeps working the same hours each week, how many weeks will it be until she earns enough money for the bike? ________

   c. What can Salena do if she wants to buy the bike in 2 weeks? ________

   ____

LESSON 10.5

Name ______________________ Date __________

Adding Decimals

Add.

1. $4.63 + 9.7$
2. $18.7 + 64.25$
3. $136.4 + 74.08$
4. $90.32 + 65.37$
5. $7.42 + 3.86$
6. $309.62 + 75.1$
7. $38.9 + 2.7$
8. $215.8 + 47.12$

9. 54.3 + 19.6 = ________
10. 5.43 + 19.6 = ________
11. 543 + 196 = ________
12. 543 + 19.6 = ________
13. 8.7 + 3.01 = ________
14. 2.79 + 45.17 = ________

Solve these problems. Show your work.

15. City workers collect trash on Marshall Street every Wednesday and Saturday. On Wednesday, they collected 343.52 kilograms of trash.

    a. On Saturday, the city workers collected 74.8 more kilograms of trash than they did on Wednesday. How many kilograms of trash did they collect on Saturday? ________

    b. How many kilograms of trash did the city workers collect altogether on Marshall Street this week? ________

16. Carmen saves 50¢ each day. How many days will it take for Carmen to save $20.00? ________

17. The distance around Pleasant Valley State Park is 68.4 miles. A park ranger drove around the park 3 times. How many miles did he drive in all? ________

LESSON 10.6

Name ______________________ Date __________

Subtracting Decimals

Subtract.

1. $\begin{array}{r} 23.7 \\ -\ 4.2 \\ \hline \end{array}$

2. $\begin{array}{r} 72.96 \\ -\ 48.02 \\ \hline \end{array}$

3. $\begin{array}{r} 55.9 \\ -\ 37.14 \\ \hline \end{array}$

4. $\begin{array}{r} 18.3 \\ -\ 6 \\ \hline \end{array}$

5. $\begin{array}{r} 7.3 \\ -\ 4.71 \\ \hline \end{array}$

6. $\begin{array}{r} 4.83 \\ -\ 4.7 \\ \hline \end{array}$

7. 34.32 − 6 = __________

8. 5.09 − 2.73 = __________

9. 509 − 273 = __________

10. 15.7 − 9.8 = __________

11. 50.9 − 2.73 = __________

12. 50.9 − 27.3 = __________

Solve these problems. Show your work.

13. Donovan bought a sports jacket that was on sale for $37.99. He gave the cashier $40.00. How much change should the cashier give Donovan? __________

14. Sanai ran 2.9 kilometers on Tuesday and 5.7 kilometers on Saturday. How many more kilometers did Sanai run on Saturday than on Tuesday? __________

15. Jasmine is 124 centimeters tall. How many meters tall is she? __________

16. Matthew earns $9.50 for each dog he walks. He walked 6 dogs on Monday and 7 dogs on Wednesday. How much money did he earn on Thursday? __________

LESSON 10.7

Name ______ Date ______

Multiplying Decimals by Whole Numbers

Multiply.

1. $\begin{array}{r} 1.05 \\ \times\ 6 \\ \hline \end{array}$
2. $\begin{array}{r} 16.33 \\ \times\ 2 \\ \hline \end{array}$
3. $\begin{array}{r} 6.2 \\ \times\ 7 \\ \hline \end{array}$
4. $\begin{array}{r} 2.05 \\ \times\ 3 \\ \hline \end{array}$
5. $\begin{array}{r} 1.73 \\ \times\ 9 \\ \hline \end{array}$
6. $\begin{array}{r} 12.4 \\ \times\ 8 \\ \hline \end{array}$
7. $\begin{array}{r} 2.38 \\ \times\ 4 \\ \hline \end{array}$
8. $\begin{array}{r} 4.1 \\ \times\ 5 \\ \hline \end{array}$

9. $8.2 \times 3 =$ ______
10. $11.6 \times 9 =$ ______
11. $2.5 \times 5 =$ ______
12. $9.03 \times 4 =$ ______
13. $7.3 \times 8 =$ ______
14. $4.83 \times 7 =$ ______
15. $8.16 \times 5 =$ ______
16. $3.58 \times 6 =$ ______
17. $2.71 \times 3 =$ ______
18. $3.20 \times 4 =$ ______
19. $32 \times 4 =$ ______
20. $3.2 \times 4 =$ ______

Solve these problems. Show your work.

21. One bag of red potatoes costs $2.59. How much do 4 bags cost? ______

22. In Mr. Beck's classroom, each desk is 65.3 centimeters long. How many centimeters long are 8 desks placed end to end? ______

LESSON 10.8

Name ______________________ Date ____________

Dividing Decimals

Solve these problems. Show your work.

1. Mrs. Gutierrez gave each of her 3 children the same amount of money to spend at the toy store. Altogether, she gave her children $25.50. How much money did each child have to spend at the toy store? ____________

2. Eight colored pencils weigh 612 grams. If each pencil has the same weight, how much does each colored pencil weigh? ____________

3. Alex drove around his block 4 times. The meter on his car said that he had driven 3.8 miles total. How many miles around is Alex's block? ____________

4. Mr. Chang built a bookcase that is 98 centimeters tall. How many centimeters long is each shelf? ____________

5. A square picture frame has a perimeter of 78.6 centimeters. How long is each side of the picture frame? Write your answer in centimeters and meters. ____________

6. Mr. Fitzgerald had 7.2 yards of fabric. Then he divided it into 3 equal pieces to make hats for the school play. How many yards was each piece? ____________

LESSON 10.9

Name ______________________ Date __________

Decimal Applications

Solve these problems. Show your work.

1. Mrs. Richards is painting her kitchen. She needs to buy 3 gallons of paint. Each gallon of paint costs $13.79. How much will the paint cost? __________

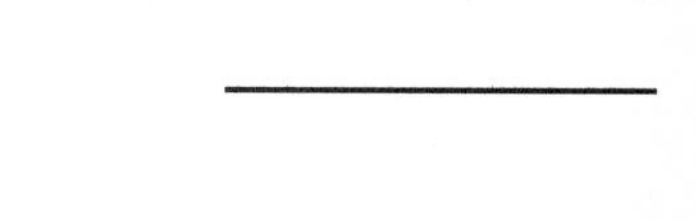

2. If Mrs. Richards gives the salesperson $50, how much change will she get? __________

Each week, Larry jogs 6.2 miles on Monday, 4.7 miles on Wednesday, and 7.5 miles on Friday.

3. How far does Larry jog each week? __________

4. How much farther does Larry jog on Friday than on Monday? __________

5. How many miles will Larry jog in 2 weeks? __________

6. Two weeks ago, Larry only jogged on Monday and Wednesday. How many miles did Larry jog that week? __________

Bill's mother deposited $8.50 in his savings account. Bill gets 50¢ allowance each week, which he deposits into his savings account. These are his balances for the past 3 weeks.

$9.00 $9.50 $10.00

7. Bill would like to know how much money he will have in his savings account 3 weeks from now. Continue the pattern for 3 more weeks.

$9.00 $9.50 $10.00 __________ __________ __________

8. How much money will Bill have 3 weeks from now? __________

LESSON 11.1

Name ______________________ Date ____________

Points and Lines

Label each of the following as a *point, line,* or *line segment.*

1. 

   ____

2. 

   ____

3. 

   ____

4. 

   ____

5. 

   ____

6. 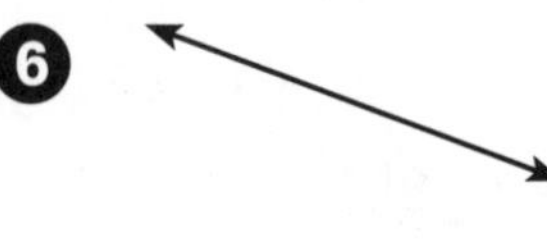

   ____

Label each pair of lines as *intersecting* or *parallel.*
If they are intersecting, point to where they will meet.

7. 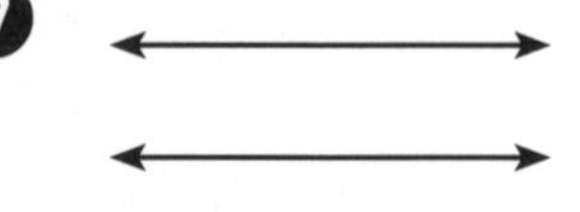

   ____

8. 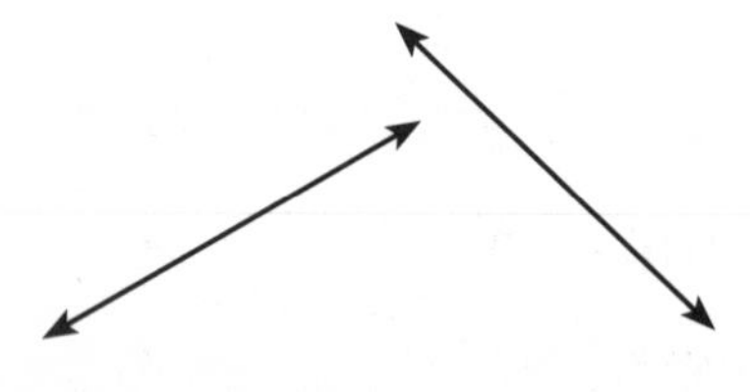

   ____

9. 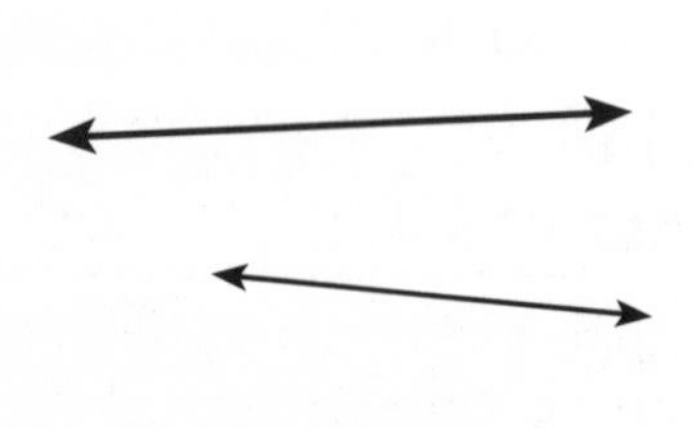

   ____

10. 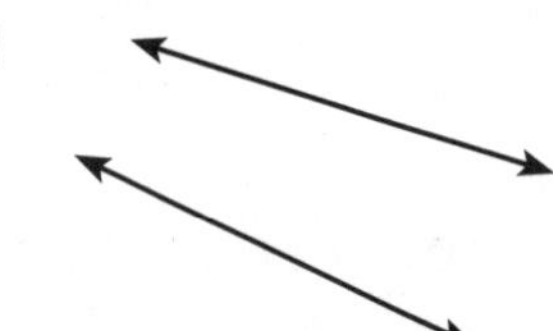

    ____

11. 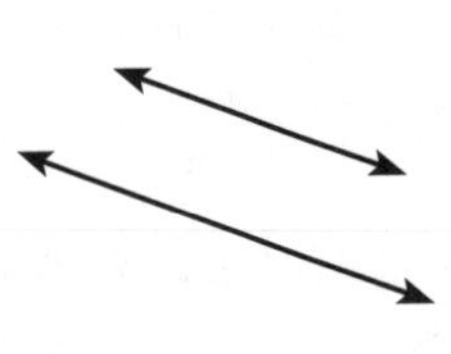

    ____

12. 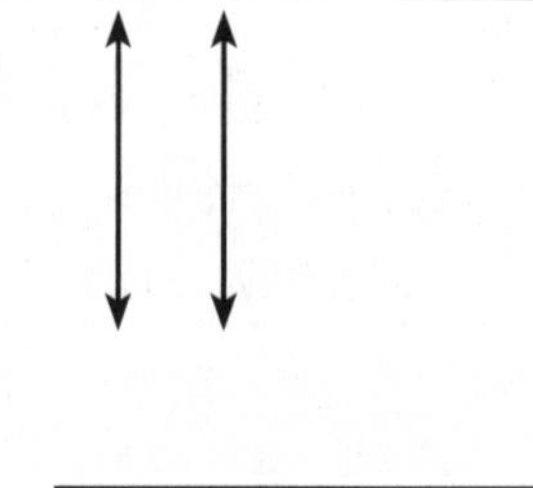

    ____

LESSON 11.2

Name ______________________ Date ____________

Angles

Label each of the following as a *ray, line,* or *angle.*

1.  ____________
2. ____________
3. ____________
4.  ____________
5. ____________
6. 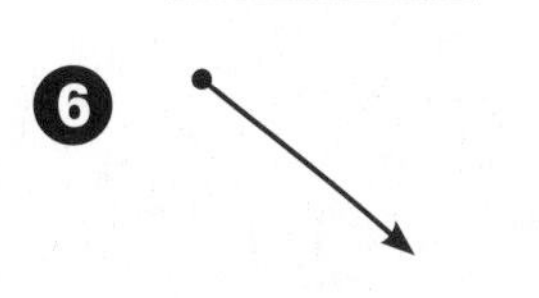 ____________

Label each angle as *acute, obtuse,* or *right.*

7. 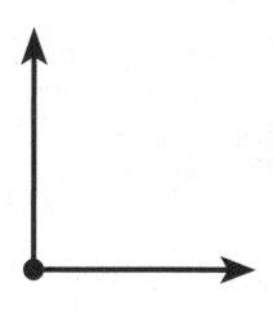 ____________

8. 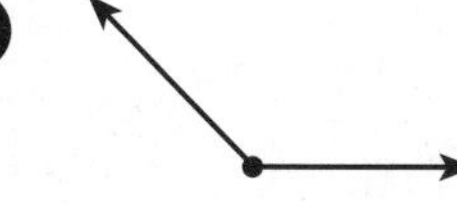 ____________
9.  ____________
10. 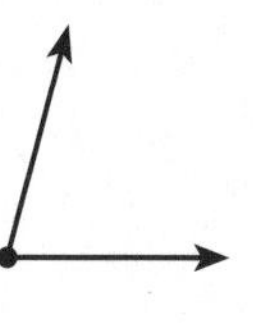 ____________
11. 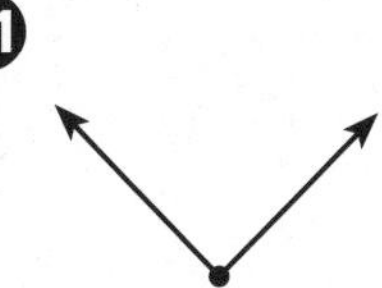 ____________
12. 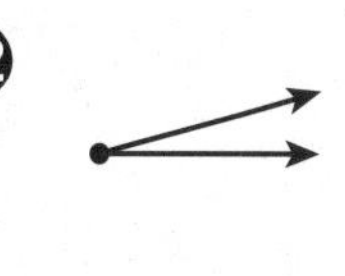 ____________

Write the correct time for each clock face. Then tell if the angle formed by the hands is *acute, obtuse,* or *right.*

13.  ____________
14.  ____________
15.  ____________
16.  ____________
17.  ____________
18.  ____________

LESSON 11.3

Name ______________________ Date __________

Triangles

Label each triangle as ***equilateral, isosceles,*** or ***right.*** **Then find the perimeter.**

1 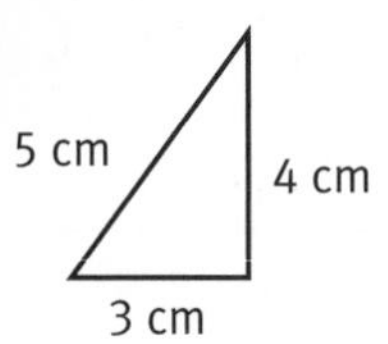

____

3 

____

2 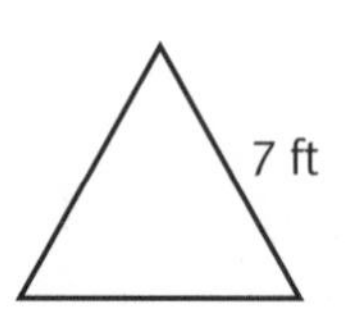

____

4 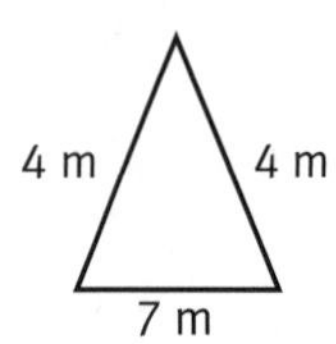

____

Ring **the correct triangles in each problem.**

5 Which of the following are right triangles?

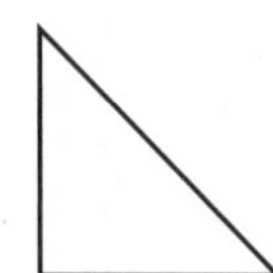  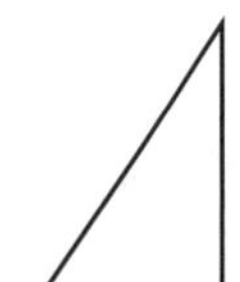 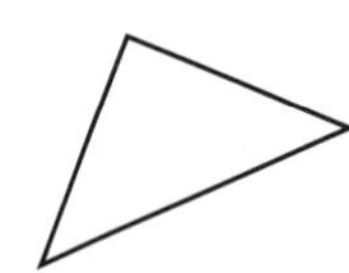 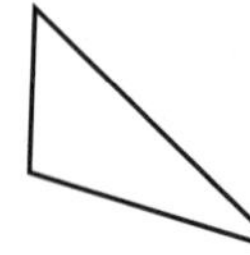

6 Which of the following are equilateral triangles?

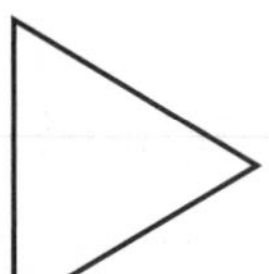 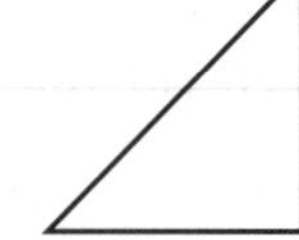 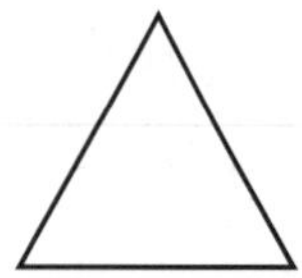 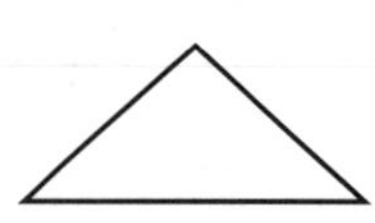 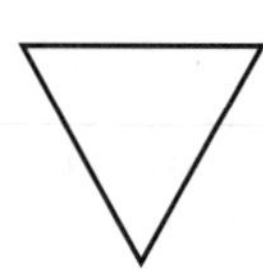

Solve.

7 Hikari's soccer team set up cones to make an equilateral triangle. A picture is shown.

10 ft

a. What is the distance around the cones? ______________

Hikari's team moved the cones to make an isosceles triangle. A picture is shown.

8 ft

4 ft

b. What is the distance around the cones? ______________

LESSON 11.4

Name ______________________ Date ____________

Quadrilaterals

Label each quadrilateral with as many names as you can.

❶ 

____

❷ 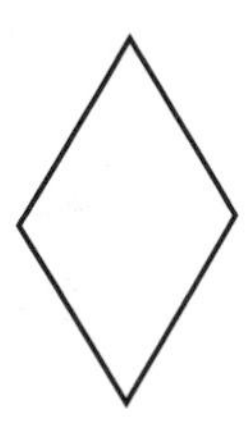

____

❸ 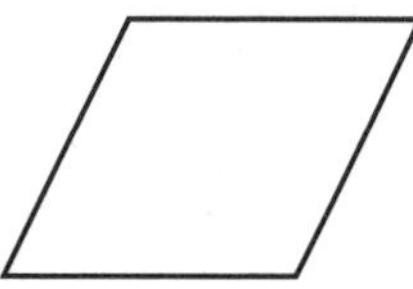

____

❹ 

____

Ring the correct figures.

❺ Which of these rectangles are squares?

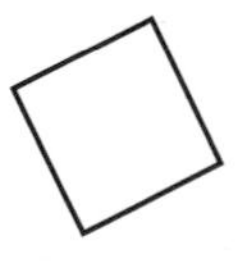   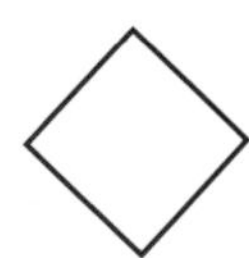  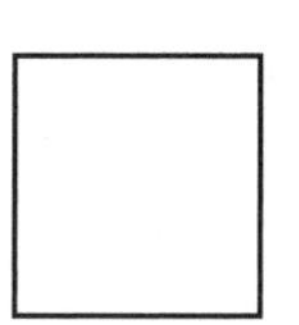 

Label each statement as True or False. Explain how you know.

❻ All trapezoids are parallelograms.

____

❼ All squares are also rectangles.

____

❽ All rhombuses are parallelograms.

____

❾ All rhombuses are squares.

____

Solve. Show your work.

❿ Sam's kite was shaped like a rhombus. One side of the kite was 2 feet. What was the perimeter of Sam's kite? ____________

LESSON 11.5

Name ______________________ Date __________

Polygons

Write *yes* or *no* to tell if each shape is a polygon. If yes, write which kind of polygon.

1. 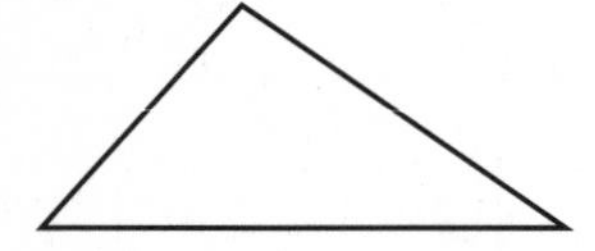
____

2. 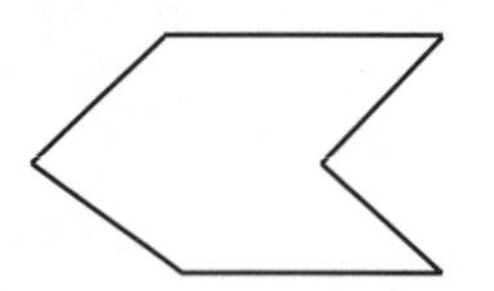
____

3. 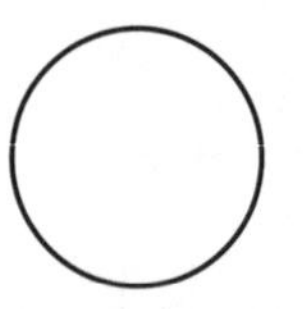
____

4. 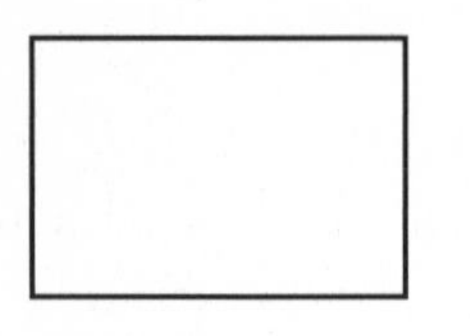
____

5. 
____

6. 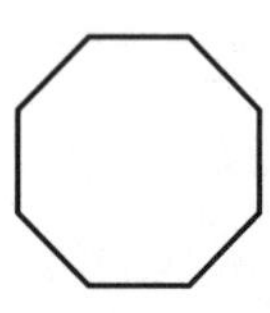
____

7. 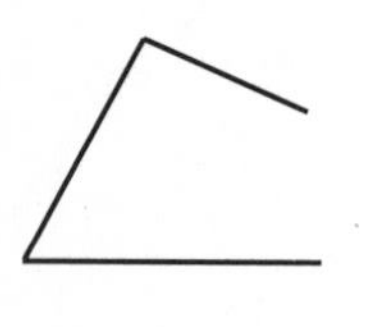
____

8. 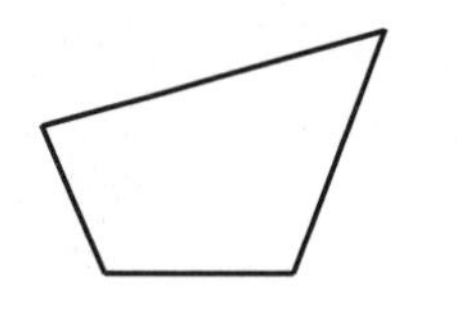
____

9. 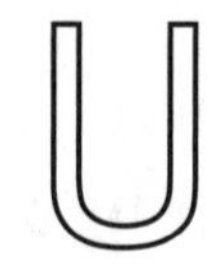
____

Draw a picture of each shape.

10. quadrilateral
11. trapezoid
12. hexagon
13. triangle

Answer each question.

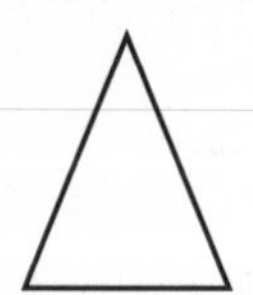

14. Number of sides? ______
15. Number of angles? ______
16. Name of shape? __________

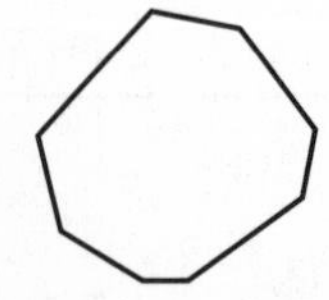

17. Number of sides? ______
18. Number of angles? ______
19. Name of shape? __________

LESSON 11.6

Name ______________________ Date __________

Congruence

Look at the pairs of figures below. Use tracing paper to help you see which are congruent. Write *yes* or *no*.

1. 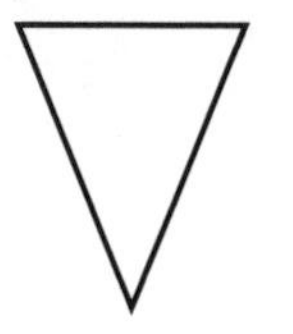

   ____

2. 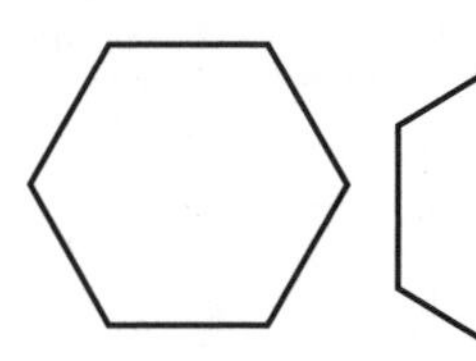

   ____

3. 

   ____

Look at the figures below. Use tracing paper to help you see which figures are congruent.

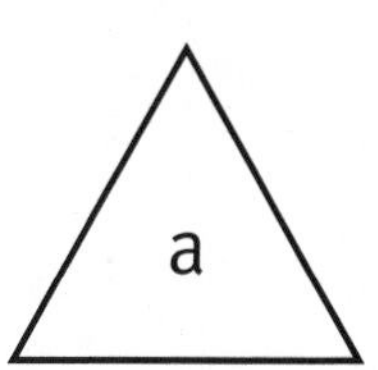

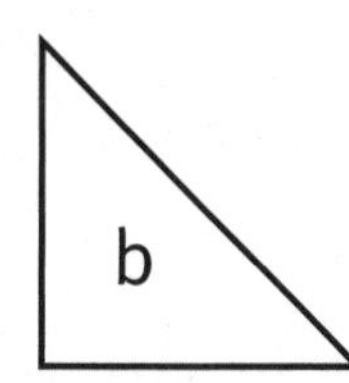

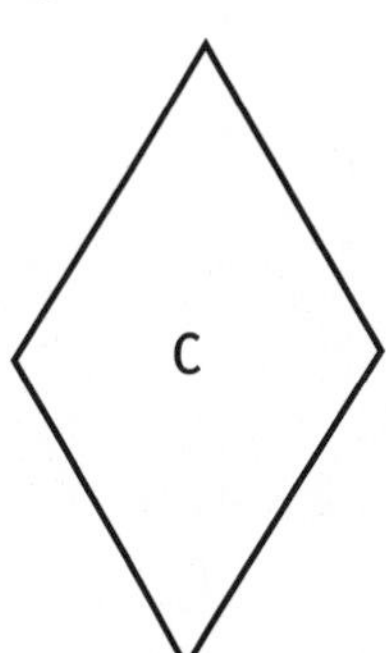

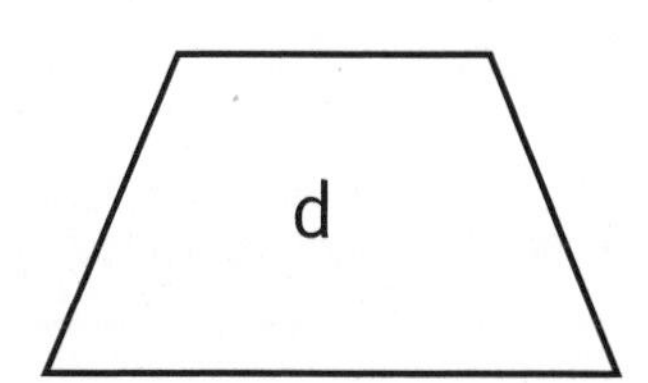

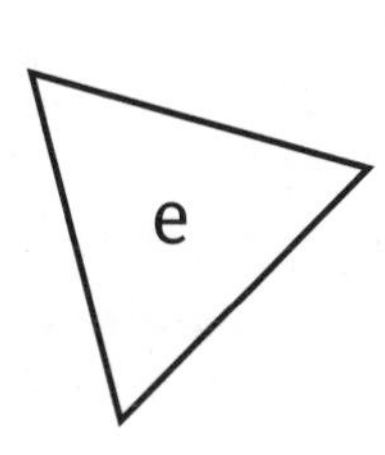

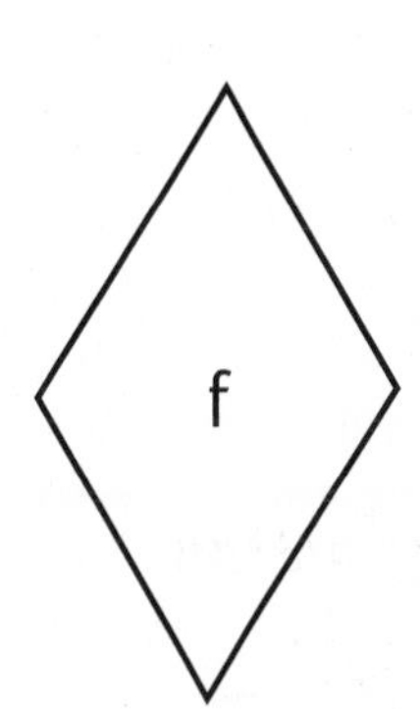

5. List the pairs of congruent figures.

   ____

6. Explain what makes two shapes congruent.

   ____

LESSON 11.7

Name ______________________________ Date ______________

Slides, Flips, and Turns

Look closely at each change. Decide whether the change was a *translation, reflection,* or *rotation*.

1 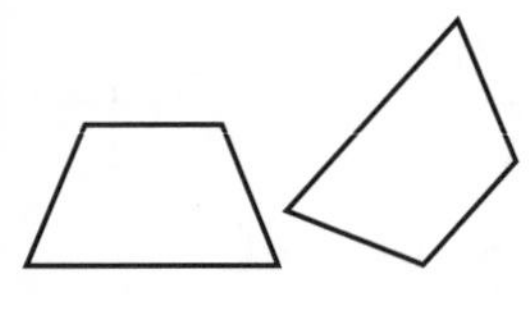

____

2 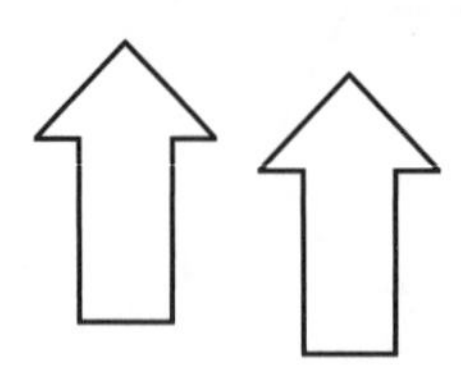

____

3 

____

4 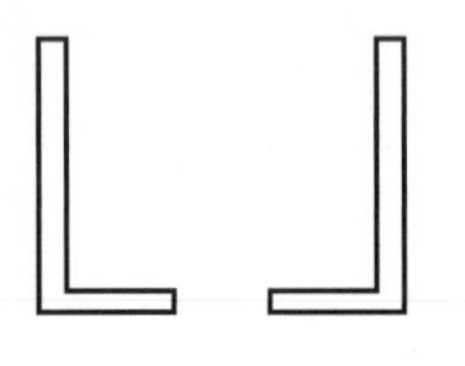

____

5 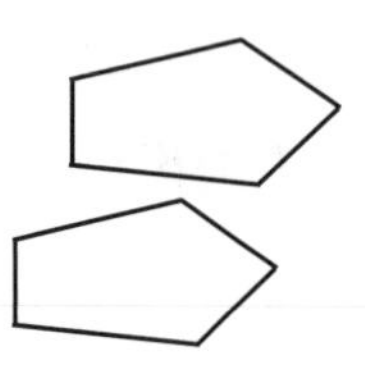

____

6 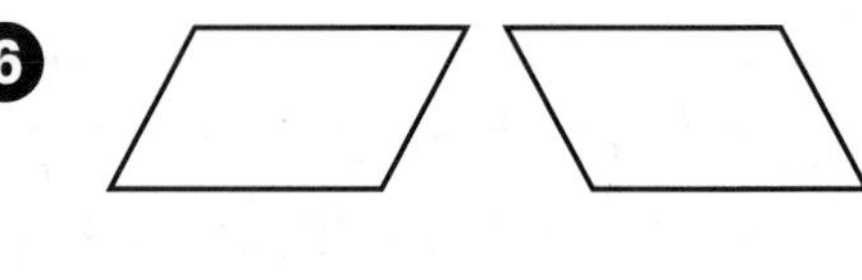

____

Describe the change from position A to position B for each figure.

7 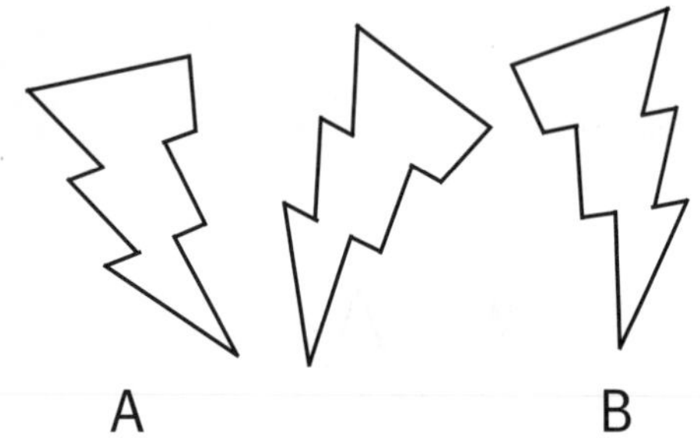

____

9 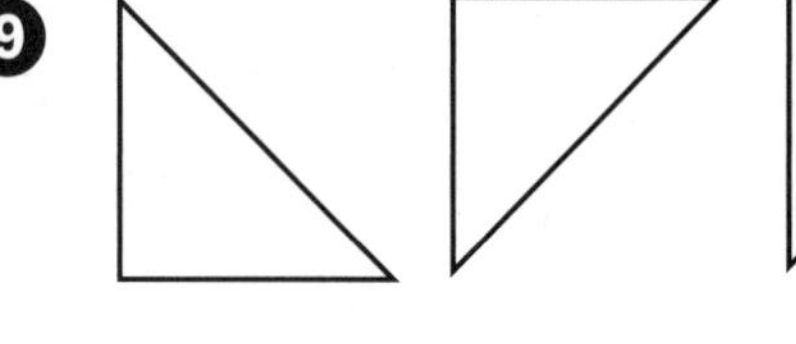

____

8 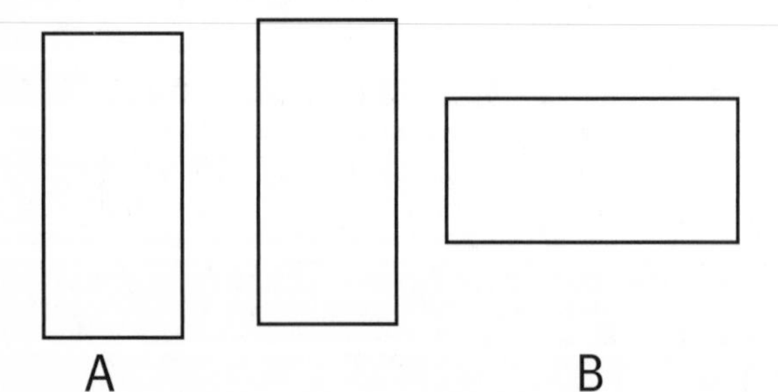

____

10 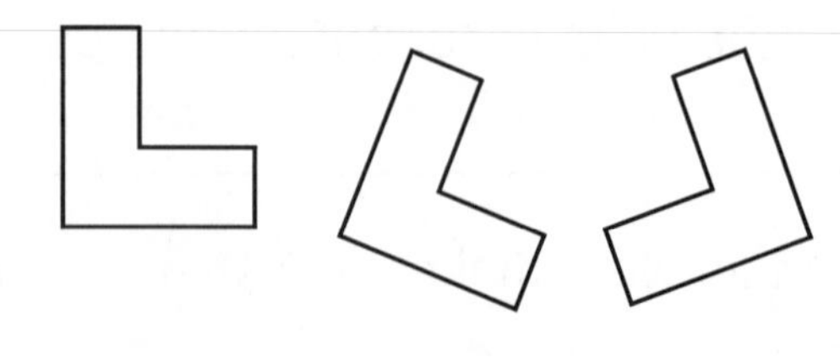

____

LESSON 11.8

Name ______________________ Date __________

Symmetry

Draw all lines of symmetry for each figure. Then write the number of lines of symmetry.

1. 
______

2. 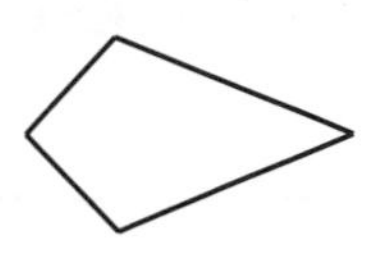
______

3. 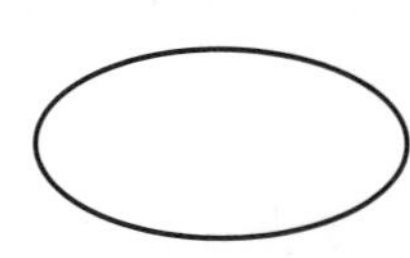
______

4. 
______

5. 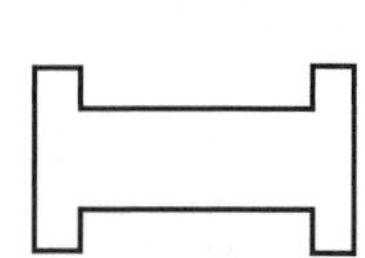
______

6. 
______

7. 
______

8. 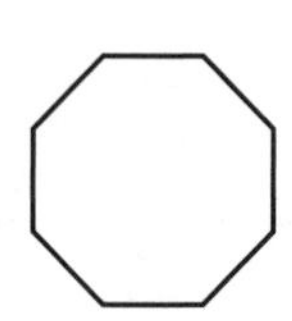
______

9. 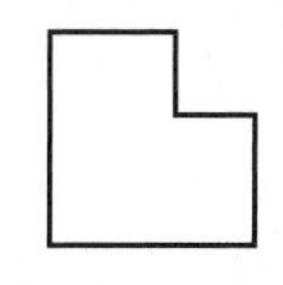
______

Is the dotted line a line of symmetry? Write *yes* or *no*.

10. 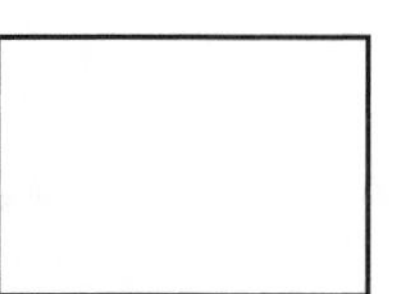
______

11. 
______

12. 
______

13. 
______

14. 
______

15. 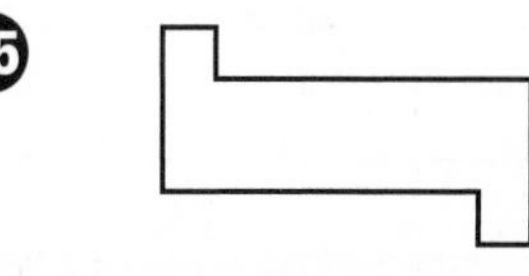
______

LESSON 11.9

Name ______________________ Date __________

Circles I

Tonya played a game of darts with LaVar.

Find the score for each dart. Measure the distance from each dart to the center. (Distances should be measured to the nearest millimeter.)

Tonya			LaVar		
Dart	Score	Distance from Center	Dart	Score	Distance from Center
A			D		
B			E		
C			F		
Total			Total		

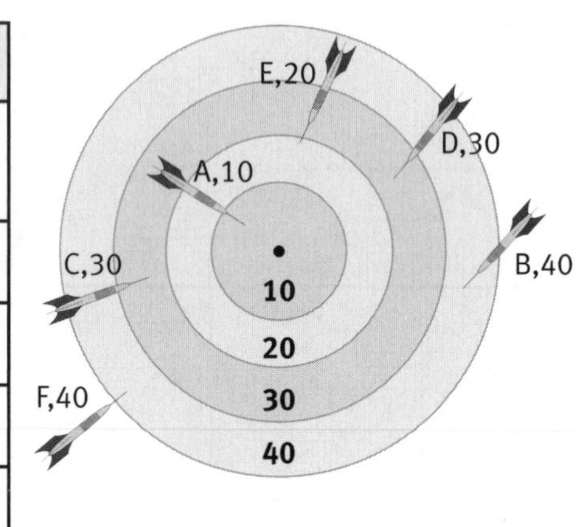

1. Who had the higher score? Explain. ______________________

Nancy and Wendy played a game of darts.

Find the score for each dart. Measure the distance from each dart to the center.

Nancy			Wendy		
Dart	Score	Distance from Center	Dart	Score	Distance from Center
A			D		
B			E		
C			F		
Total			Total		

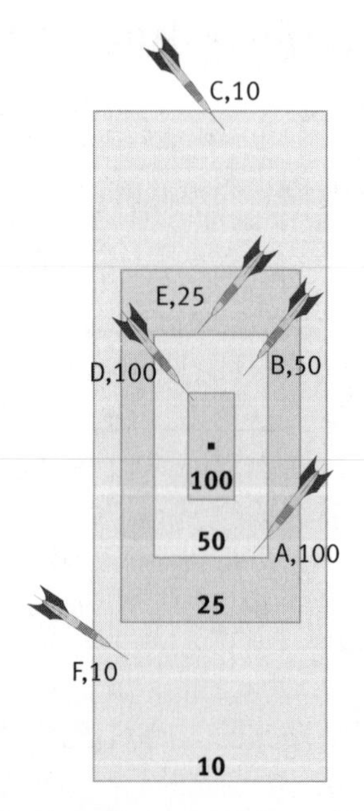

2. Who had the higher total score? __________
3. Which player do you think had more skill? Explain.

____

4. Which dartboard seems more fair, the circular or the rectangular dartboard?

____

LESSON 11.10

Name ______________________ Date __________

Circles II

Use Circle *X* to solve the following.

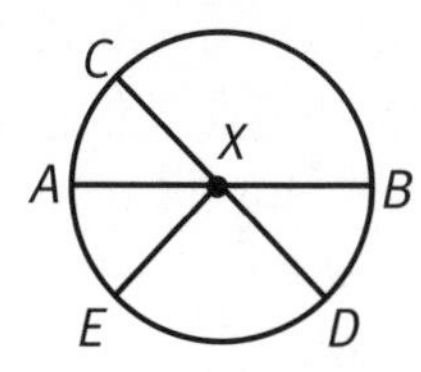

1. Radius *XE* is ______ centimeters long.
2. Diameter *AB* is ______ centimeters long.
3. Radius *XA* is ______ centimeters long.
4. Diameter *CD* is ______ centimeters long.

Use Circle *R* to solve the following.

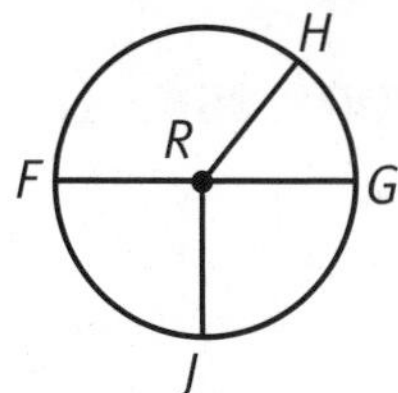

5. Name the circle. ______________
6. Name a radius. ______________
7. Name a diameter. ______________

Use Circle *M* to identify the following statements as either *true* or *false*.

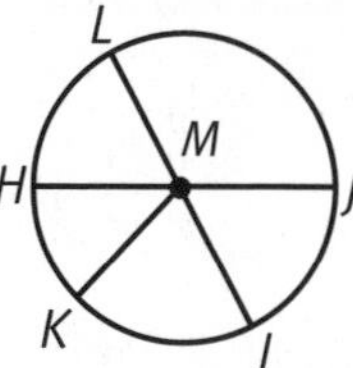

8. Line Segment *MH* is a diameter. ____________
9. Line Segment *LI* is a diameter. ____________
10. Line Segment *HJ* is a radius. ____________
11. Line Segment *MK* is a radius. ____________
12. Any diameter is twice as long as any radius. ____________

LESSON 11.11

Name ______________________ Date __________

Space Figures

Answer the following questions about each figure.

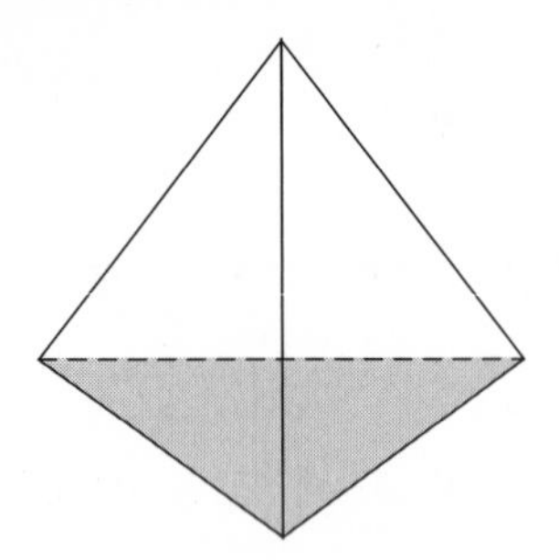

1. How many faces? ______
2. How many edges? ______
3. How many vertices? ______

4. How many faces? ______
5. How many edges? ______
6. How many vertices? ______

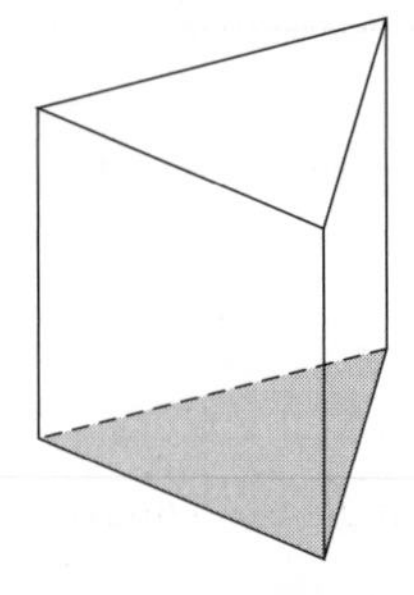

7. How many faces? ______
8. How many edges? ______
9. How many vertices? ______

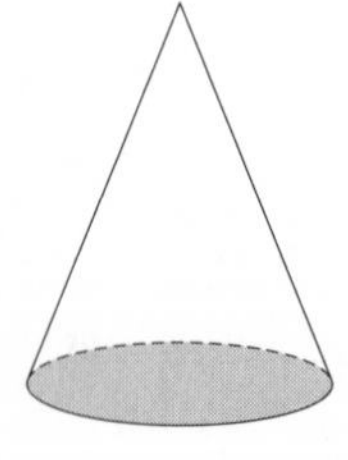

10. How many faces? ______
11. How many edges? ______
12. How many vertices? ______

LESSON 11.12

Name ______________________ Date __________

Nets and Surface Area

Find the area of the figures below. Each small square has an area of one square centimeter.

❶ 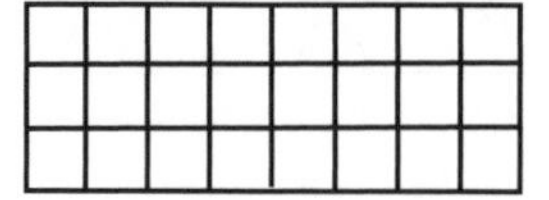

_______ square centimeters

❷ 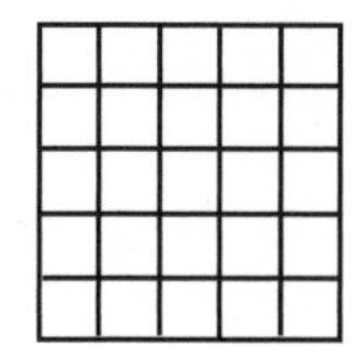

_______ square centimeters

Look at the nets of each prism. Use the nets to answer each question.

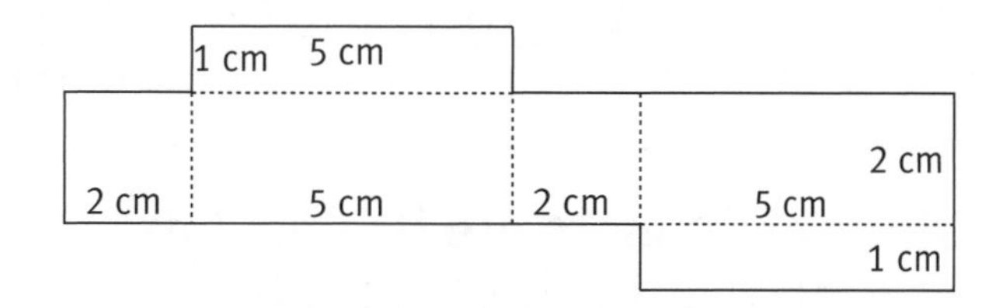

❸ What are the shapes of the polygons in the net? ______________

❹ How many polygons are in the net? ________

❺ What is the area of each polygon in the net? ______________

❻ What is the surface area of the whole net? ______________

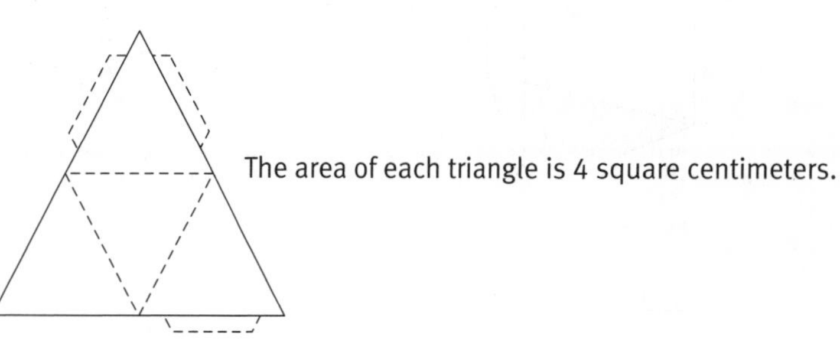

❼ What are the shapes of the polygons in the net? ______________

❽ How many polygons are in the net? ________

❾ What is the area of each polygon in the net? ______________

❿ What is the surface area of the whole net? ______________

LESSON 11.13

Name ______________________ Date ____________

Volume

What is the volume? Count the cubes to find the volume.

1. 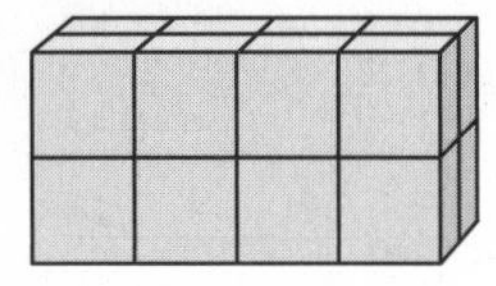

_______ cubic units

2. 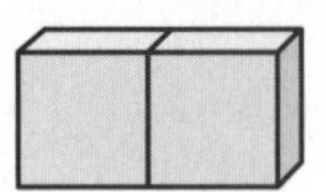

_______ cubic units

3. 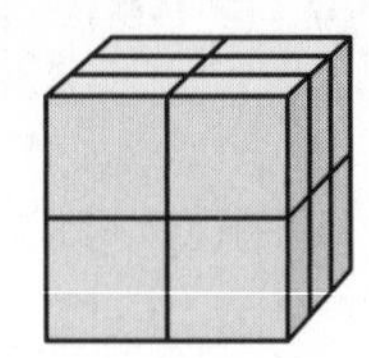

_______ cubic units

4. 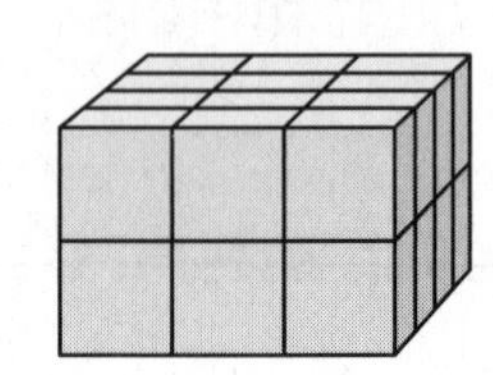

_______ cubic units

Find out how many cubic centimeters are in each box. Then give the volume of each box. Each small cube is one cubic centimeter.

5. 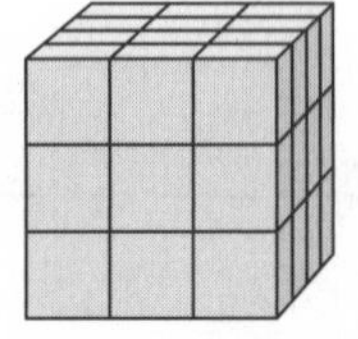

_______ cubic centimeters

6. 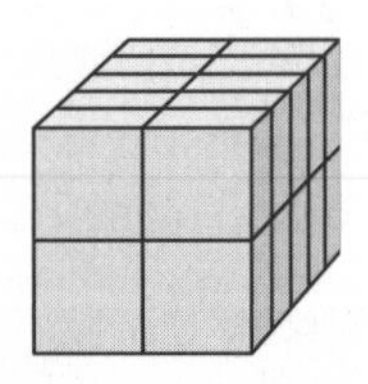

_______ cubic centimeters

7. 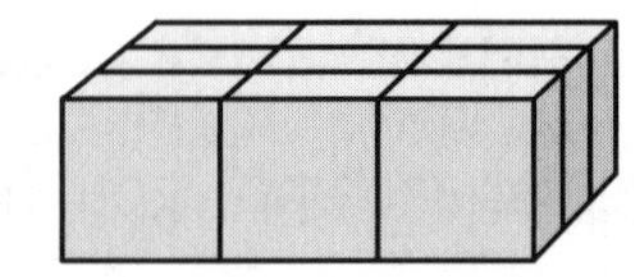

_______ cubic centimeters

8. 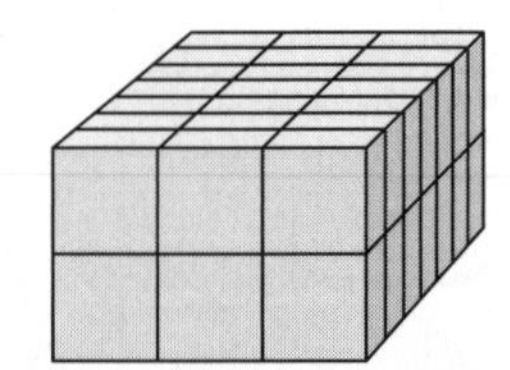

_______ cubic centimeters

Write the equivalent amount.

9. 3 L = _______ mL
10. 5 L = _______ mL
11. _______ L = 2,000 mL
12. _______ L = 9,000 mL

Name ______________________ Date ____________

Collecting Data—Samples

Decide **if each of the following shows a good sample by answering *yes* or *no*. If it is not a good sample, explain why not.**

1. To sample the students at Roosevelt Elementary School, the third-grade class asked questions of the girls.

   ____

2. To survey the customers of a local music store, Mr. Bing's music class asked questions of one customer every 15 minutes for 8 hours.

   ____

Mrs. Edwards' third-grade class did research to find out how fast different animals could run. The class recorded their results.

giraffe—35 miles per hour

zebra—40 miles per hour

cheetah—70 miles per hour

elephant—15 miles per hour

feral pig—30 miles per hour

bear—25 miles per hour

Create **a table for the information above in the space provided.**

3. Write one true and one false statement about the information in the table.

   ____

LESSON 12.2

Name ____________________ Date __________

Collecting Data—Tally Marks

José surveyed the children in his neighborhood to find out what kind of movie they like. He recorded the following results.

Movie	Tally	Number of Children
Comedy	𝍸	5
Action	𝍸 ||	7
Animation	|||	3
Science Fiction	𝍸 𝍸 ||||	14
Mystery	𝍸 |	6
	Total	35

Use the table to answer the following questions.

1. What is the most common favorite kind of movie in the survey?

   ____

2. Do you think this is the favorite kind of movie of the entire neighborhood? Why or why not? ____________________

   ____

3. What would be a good title for this survey? ____________________

Use the following information to fill in the table below.

Kerry did a survey to see where she would find the most ladybugs in her backyard. Kerry found 10 ladybugs on a window, 9 ladybugs in the grass, 3 ladybugs by a fence, and 16 ladybugs by a tree.

Place Found	Tally	Frequency
	𝍸 ||||	
Tree	𝍸 𝍸 𝍸 |	
	𝍸 𝍸	10
	|||	
Other	𝍸 ||	7
	Total	

Answer the following questions about the completed table from Kerry's survey.

4. Did Kerry find more ladybugs on a window or by a tree?

   ____

5. Write one true and one false statement about the information you just charted.

   ____

LESSON 12.3

Name ______________________ Date __________

Summarizing Data

Mr. Albertson took his students to a berry farm to pick blackberries. He recorded the results in the following chart.

Student Name	Number of Berries Picked
Olga	34
Pawel	61
Markus	35
Whitney	46
Kendra	39
Bryan	33
Carlos	42

Use the chart to answer the following questions.

1. What is the least number of berries picked by a student? The greatest number? __________

2. What is the range for the number of berries picked? __________

3. Which number might be considered to be the *outlier*? __________

4. True or False. The number of berries picked varied greatly from student to student. __________

Hour	Number of Cars
1	18
2	15
3	12
4	3
5	10
6	15

Mount Eagle Elementary School raised money to build a new library by having a car wash. The table shows how many cars were washed during each hour of one day.

Use the chart to answer the following questions.

5. What is the least number of cars washed? The greatest number? __________

6. Which hour might be considered an outlier? __________

7. What might be the reason for the outlier? ______________________

LESSON 12.4

Name ______________________ Date __________

Mean, Median, Mode

A basketball coach measured the height of each player on the team. Here are the heights in inches of all the players.

71	74	69
67	82	71
75	71	77

Use the chart of heights of the basketball players to answer the following questions.

1. List the numbers in order from least to greatest. ______________________
2. What is the least number in your list? The greatest? __________
3. Which numbers appear most often (the *mode*)? __________
4. What is the *mean* or average height for the players (in inches)? __________

The water level in a swimming pool was recorded several times in June. Here are the results in gallons.

250	300	350	400
200	250	450	300
450	250	350	300

Use the chart of the water levels to answer the following questions.

5. What is the *mode* for these water levels? __________
6. What is the average water level (the *mean*)? __________
7. What are the lowest and highest water levels? __________
8. What amount is the *median* or middle water level? __________

Name ______________________ Date ____________

Displaying Data

Jamie is saving money for her summer vacation. She recorded the amount saved each month in a notebook.

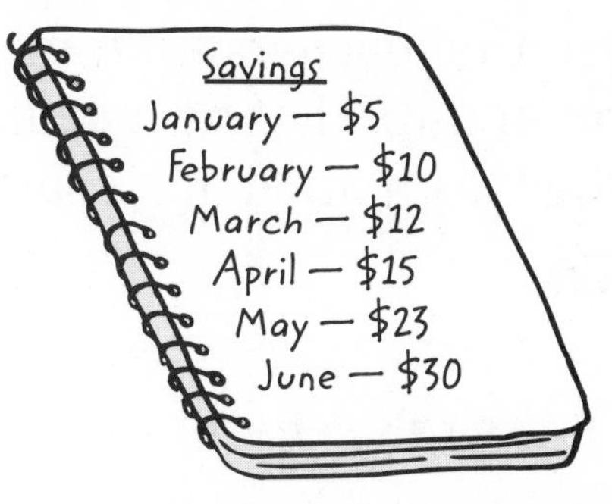

Use the information from Jamie's notebook to complete the table.

Month	Amount Saved
January	$5
March	
April	$15

Answer the following questions.

1. How much money did Jamie save in May? __________
2. Jamie saved $15 in which month? __________
3. How much more did Jamie save in June than March? __________

The grid shows a section of a local zoo. Each location can be named with an *ordered pair.*

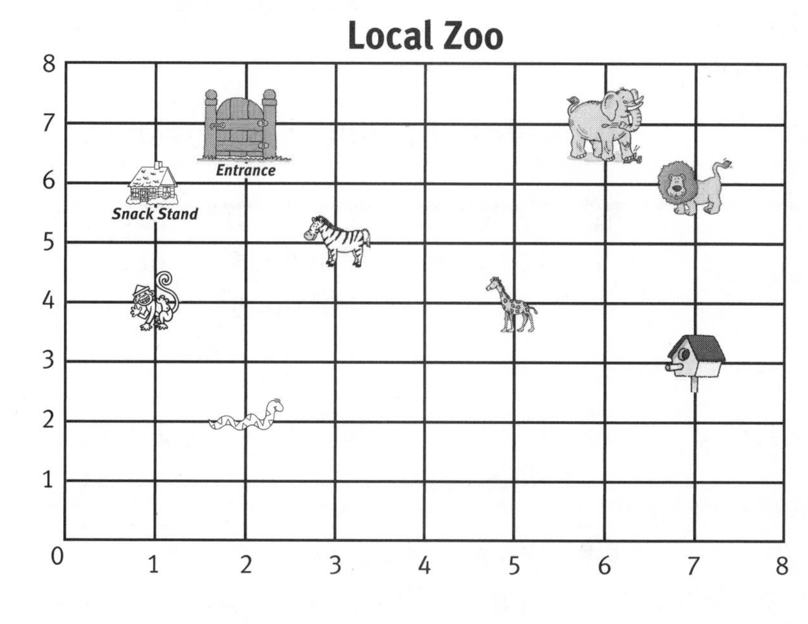

Answer the following questions by using the grid.

4. Where is the birdhouse located? __________
5. Where is the entrance located? __________
6. Which animals are farthest west, the lions or the monkeys? __________
7. If you were traveling from the elephants to the snakes, would you travel north or south? __________

Name ____________________ Date __________

Graphs That Compare

Obi surveyed the teachers at Tremont Elementary School to find out their favorite kind of dog. First he created a tally chart to show the results. After recording the information in a chart, Obi created a bar graph.

Complete the tally chart.

1.

Dog	Tally	Frequency
Beagle	||	
Poodle	𝍸 𝍸 ||	
Dalmatian	|	
Shepherd	𝍸 |||	
Other	𝍸 ||	
	Total	

Use the information from the tally chart to complete the bar graph.

Kind of Dog

Obi also created a circle graph to show the percentages for each favorite dog.

Complete the circle graph.

Answer the following questions.

3. Which appear to be the most common and the least common favorite kinds of dogs?

____

4. How many teachers were surveyed altogether? ____________________

5. Write a sentence describing the information in each graph.

____

____

LESSON 12.7

Name ____________________ Date __________

Graphs Showing Change

This line graph shows the change in movie ticket prices through the years.

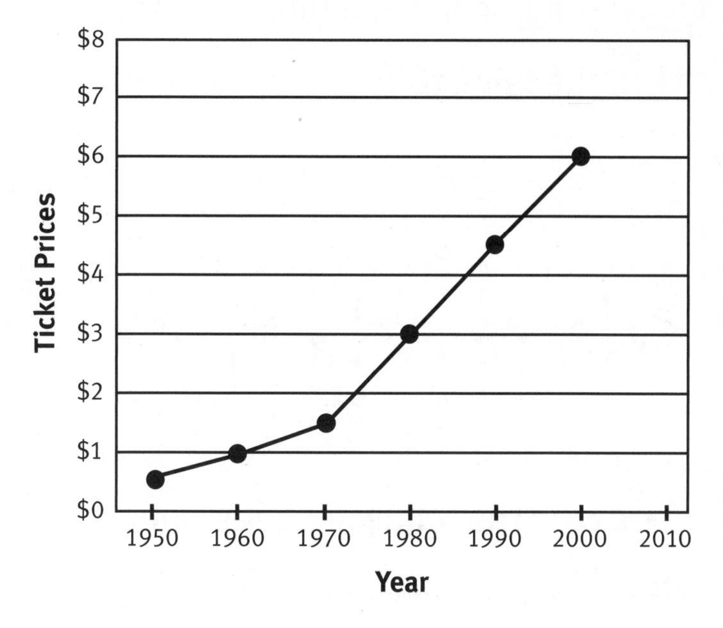

Use the graph to answer these questions.

❶ What would be a good title for the graph? ____________________

❷ Does this line graph show an *increase* or a *decrease* in ticket prices? __________

What were the ticket prices in

❸ 1960? ______ ❹ 1990? ______

What year was it when ticket prices were

❺ $1.50? __________ ❻ $3.00? __________

About how much did ticket prices increase between

❼ 1950 and 1960? ________ ❽ 1990 and 2000? ________

❾ If tickets continue to increase at the same rate as they did between 1970 and 1980, how much do you think movie tickets will be in the year 2010? ________

Use the graph to answer these questions.

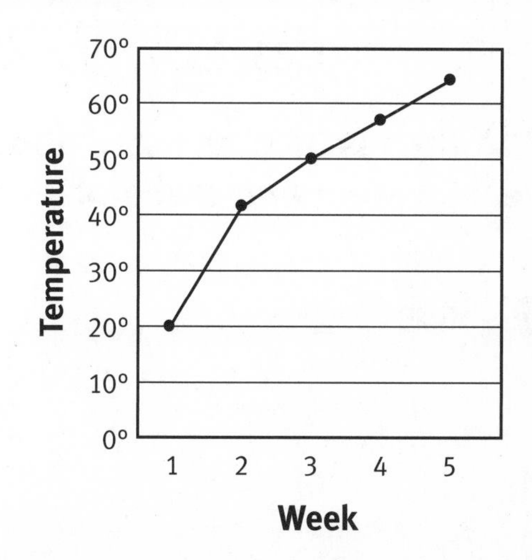

❿ What would be a good title for the graph?

____

⓫ Does this line graph show an *increase* or a *decrease* in the temperature? __________

LESSON 12.8

Name ______________________________ Date ______________

Graphs Showing How Data Is Grouped

Mrs. Elliott's science class planted sunflower seeds. After 10 weeks, Mrs. Elliott recorded the height of each child's sunflower in a table.

Student Name	Sunflower Height in Inches
Amber	7
Kaitlin	9
Anthony	10
Alex	8
Michelle	9
Stephanie	5
Francesco	8
Reynaldo	9
Rosie	7
Orna	10
Jamal	10
Dominck	8
Julie	8
Manuel	9
Jerome	8

Use the table to create a line plot.

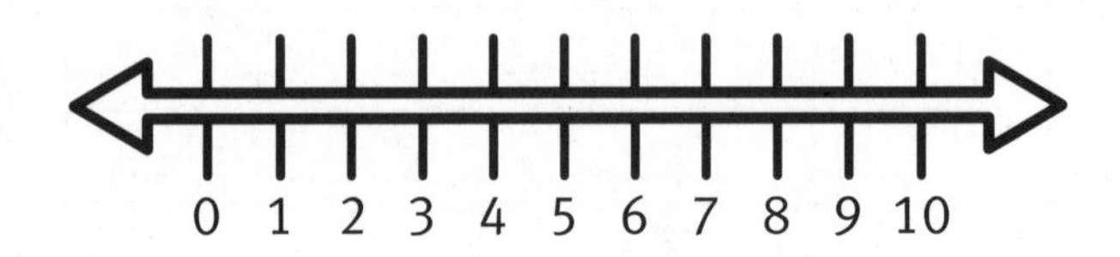

Answer the following questions.

2. What is the mode? __________
   What is the range? __________

3. Which height might be considered an *outlier?* __________

4. Write one true statement about the graph.

   ____

The following are math test scores for the students in Mr. Kammer's class.

89, 85, 73, 80, 85, 93, 77, 90, 82, 87, 85, 82, 95, 75, 84, 86

Use the test scores listed above to create a stem-and-leaf plot.

5. Stem | Leaves

Answer the following questions.

6. How many students scored between an 80 and 89? __________

7. Write one true and one false statement about the stem-and-leaf plot.

   ____

   ____

LESSON 12.9

Name ______________________ Date ____________

Interpreting and Analyzing Data

Orion is about to run a 5 kilometer race. He has practiced five hours a week for eight weeks before the race. He has created a *scatter plot* to see if his running averages have improved over time.

Answer the following questions.

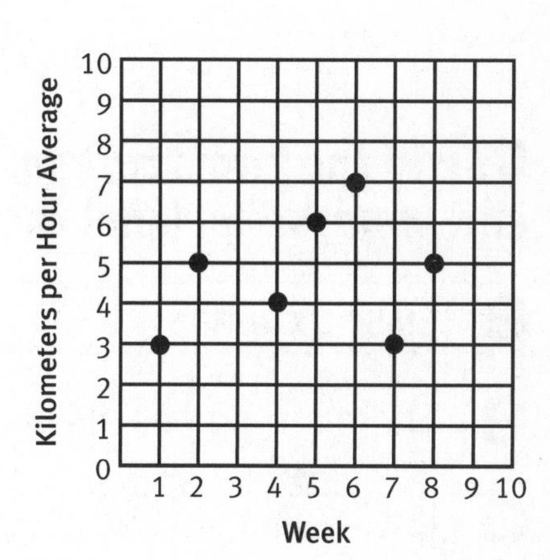

1. Did Orion's running time get better or worse overall? ______________________

   ____

2. True or False: this graph resembles a line.

   ____

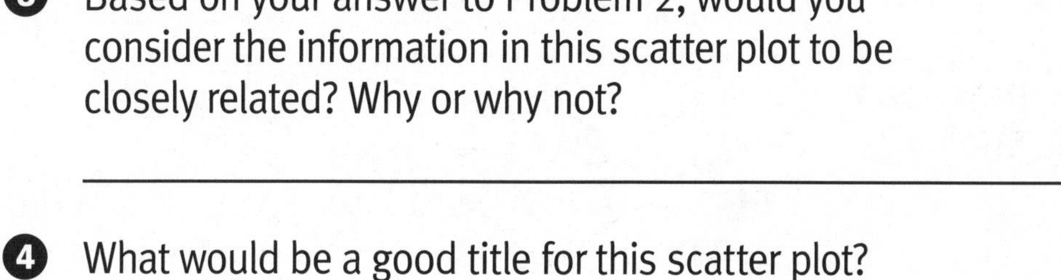

3. Based on your answer to Problem 2, would you consider the information in this scatter plot to be closely related? Why or why not?

   ____

4. What would be a good title for this scatter plot?

   ____

Justin surveyed several students at Hillside Elementary School to find out which season was the favorite. Here are his results: winter had 21 votes, spring had 37 votes, summer had 55 votes, and fall had 32 votes.

Use the information Justin gathered to create a *pictograph* and a *bar graph* on a separate sheet of paper.

5. What is a good title for the graphs? ____________

6. How many students were surveyed in all? ____________

7. Write one true statement about one of the graphs.

   ____

LESSON 12.10

Name ______________________ Date ____________

Probability—Impossible to Certain

0 ______________________________ 1

Impossible — Unlikely — Equally Likely — Likely — Certain

Using the terms *certain, likely, equally likely, unlikely,* and *impossible,* label each of the following scenarios:

1. Jesse's car has 4 tires. ____________
2. There are no pencils in Beacon Heights Elementary School. ____________
3. It will rain cats and dogs today. ____________

Answer the following questions based on the marbles in the bag.

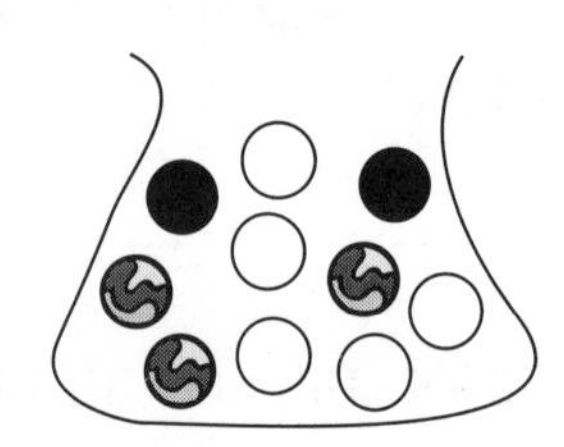

4. What is the probability (as a fraction) of choosing a black marble? ________
5. What is the probability (as a fraction) of choosing a white marble? ________
6. What is the probability (as a fraction) of choosing a striped marble? ________

There are 8 yellow counters and 4 orange counters in a bag.

7. What is the probability (as a fraction) of choosing a yellow counter? ________
8. What is the probability (as a fraction) of choosing an orange counter? ________

LESSON 12.11

Name ______________________ Date ____________

Probability—Predictions and Experiments

Use the spinner to assist with the following questions.

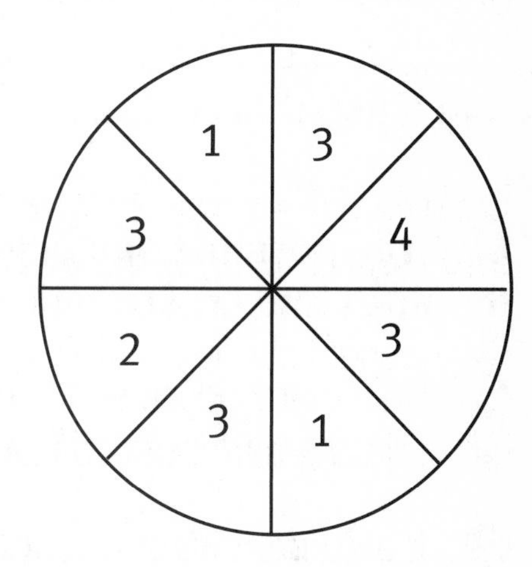

1. What is the probability (as a fraction) that you will spin a 1? A 2? A 3?

____

2. If you were to spin the spinner 4 times, what is the number of times that you would expect to spin a 1?

____

3. If you were to spin the spinner 16 times, what is the number of times that you would expect to spin a 3?

____

4. Is the spinner more likely to land on a 4 or a 3? Explain your thinking.

____

____

Suppose you roll two 0 – 5 *Number Cubes*.

5. Copy and complete the chart, using your knowledge of sums.

	0	1	2	3	4	5
0	0	1	2	3	4	5
1	1	2	3	4	5	6
2	2				6	7
3	3	4		6		8
4	4	5				9
5	5					10

6. What is the probability (as a fraction) that you would roll a sum of 9?

____

7. What is the probability (as a fraction) that you would roll a sum of 15?

____

8. What is the probability (as a fraction) that you would roll a sum of 4?

____

Name ______________________ Date ____________

Displaying and Analyzing Outcomes

The cook at Pierce Elementary School needed to prepare pizza lunches for the students. She made a table to show all the combinations (possible outcomes) for pizza toppings and crust. Students will have a choice of cheese, chicken, mushroom, thin crust, thick crust, and stuffed crust.

Topping	Crust
Chicken	Thin
Cheese	Thin
Mushroom	

List the possible outcomes for the pizza toppings and crust combinations.

1. How many possible outcomes are there for pizza topping and crust combinations? ________

2. If a student wants a cheese pizza, how many possible outcomes are there? ________

Natasha has to choose an outfit to wear to a concert. She has a skirt, a pair of pants, a blue sweater, a black sweater, a pair of boots, and a pair of shoes in her closet.

Complete the following tree diagram using the information about Natasha's clothing choices.

5.

skirt
- ____________
  - boots
  - shoes
- black sweater
  - ____________
  - ____________

pants
- blue sweater
  - ____________
  - shoes
- ____________
  - boots
  - ____________

Use the completed diagram to answer the following questions.

6. How many possible outcomes are there? ________

7. If Natasha wants to wear a blue sweater, how many possible outcomes are there? ________

Name ______________________ Date __________

100 Table

0	1	2	3	4	5	6	7	8	9
10	11	12	13	14	15	16	17	18	19
20	21	22	23	24	25	26	27	28	29
30	31	32	33	34	35	36	37	38	39
40	41	42	43	44	45	46	47	48	49
50	51	52	53	54	55	56	57	58	59
60	61	62	63	64	65	66	67	68	69
70	71	72	73	74	75	76	77	78	79
80	81	82	83	84	85	86	87	88	89
90	91	92	93	94	95	96	97	98	99

MASTERS

Name ______________________ Date ____________

The Multiplication Table

×	0	1	2	3	4	5	6	7	8	9	10
0	0	0	0	0	0	0	0	0	0	0	0
1	0	1	2	3	4	5	6	7	8	9	10
2	0	2	4	6	8	10	12	14	16	18	20
3	0	3	6	9	12	15	18	21	24	27	30
4	0	4	8	12	16	20	24	28	32	36	40
5	0	5	10	15	20	25	30	35	40	45	50
6	0	6	12	18	24	30	36	42	48	54	60
7	0	7	14	21	28	35	42	49	56	63	70
8	0	8	16	24	32	40	48	56	64	72	80
9	0	9	18	27	36	45	54	63	72	81	90
10	0	10	20	30	40	50	60	70	80	90	100

MASTERS

Name ______________________ Date ____________

Tenths and Hundredths Circle

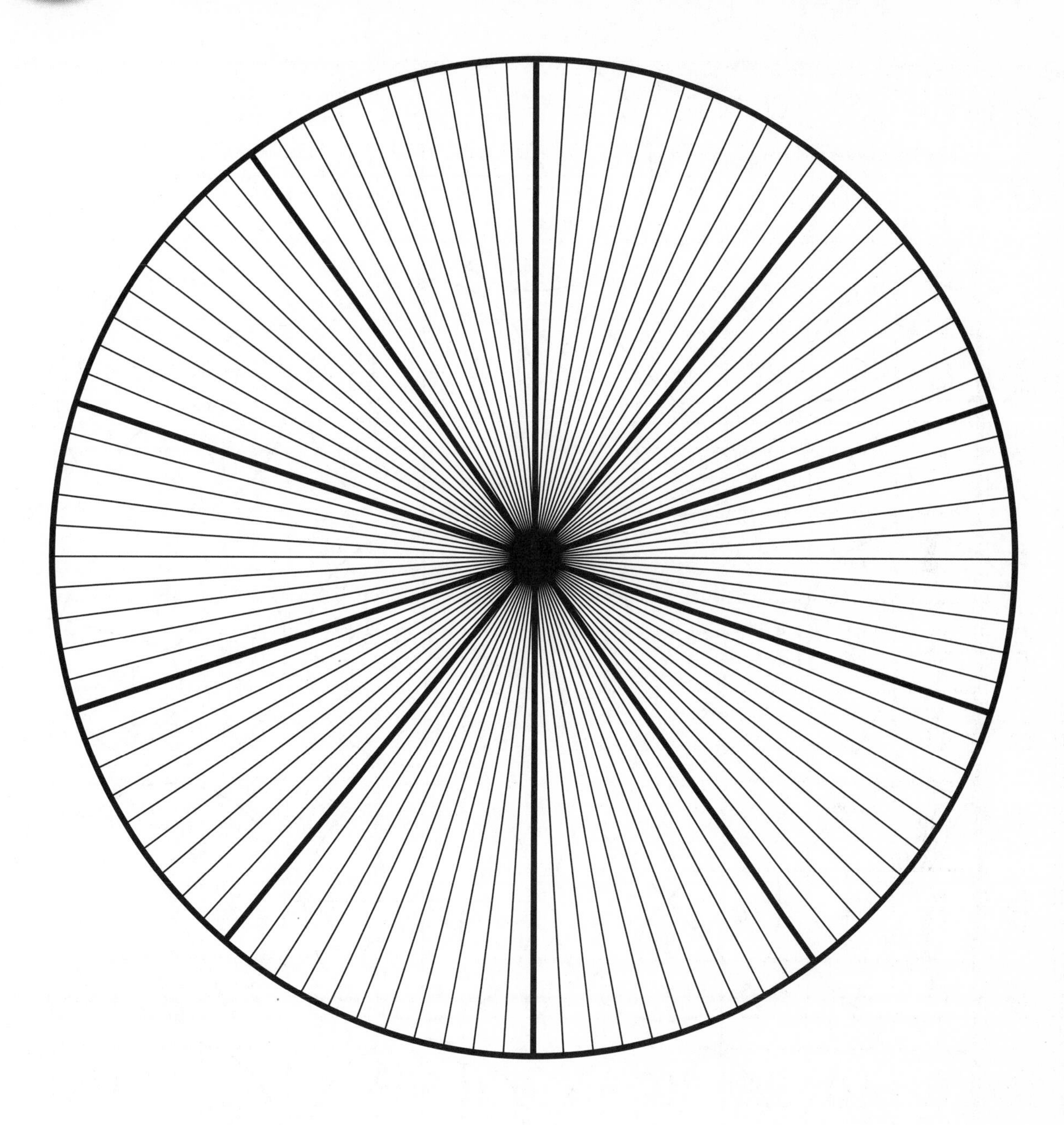

MASTERS

Name ____________________ Date __________

Fraction Game

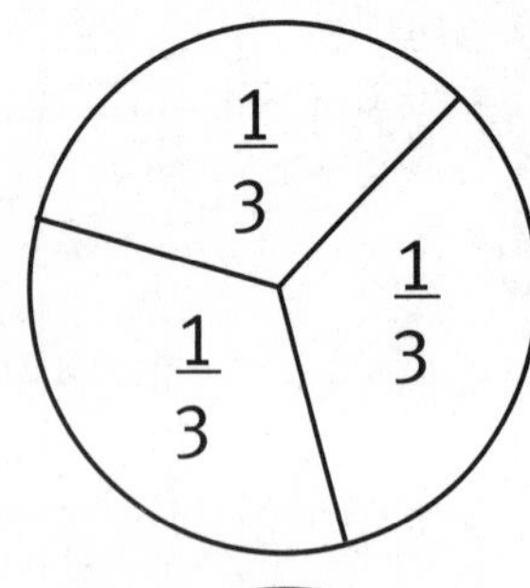

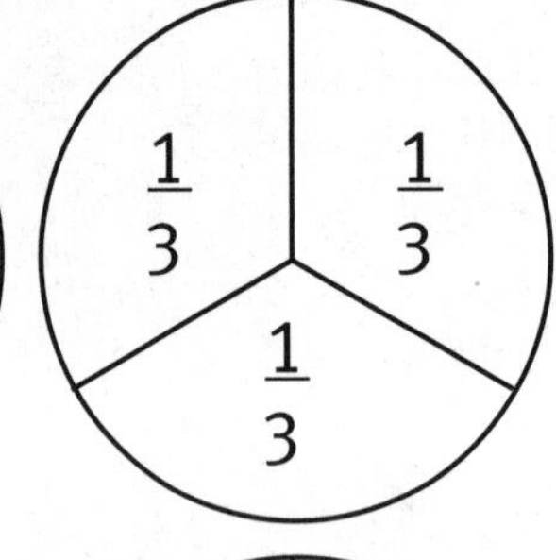

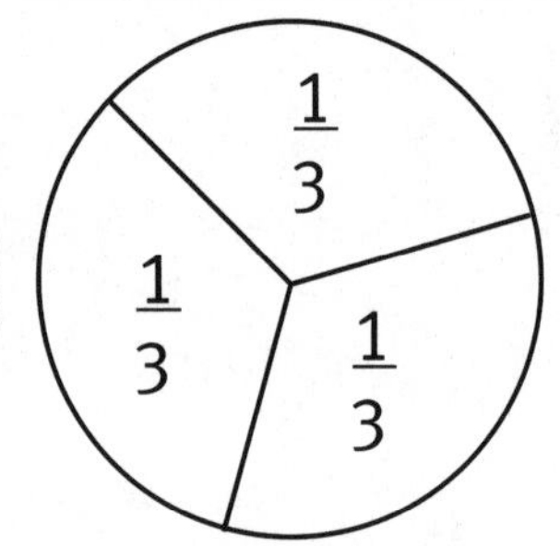

MASTERS

Name ______________________ Date ____________

Graph Town

N
E W
S

10th Avenue
9th Avenue
8th Avenue
7th Avenue
6th Avenue
5th Avenue
4th Avenue
3rd Avenue
2nd Avenue
1st Avenue
0th Avenue

D
C
B
A

0th Street
1st Street
2nd Street
3rd Street
4th Street
5th Street
6th Street
7th Street
8th Street
9th Street
10th Street

MASTERS

Name ______________________ Date ____________

Centimeter Graph Paper

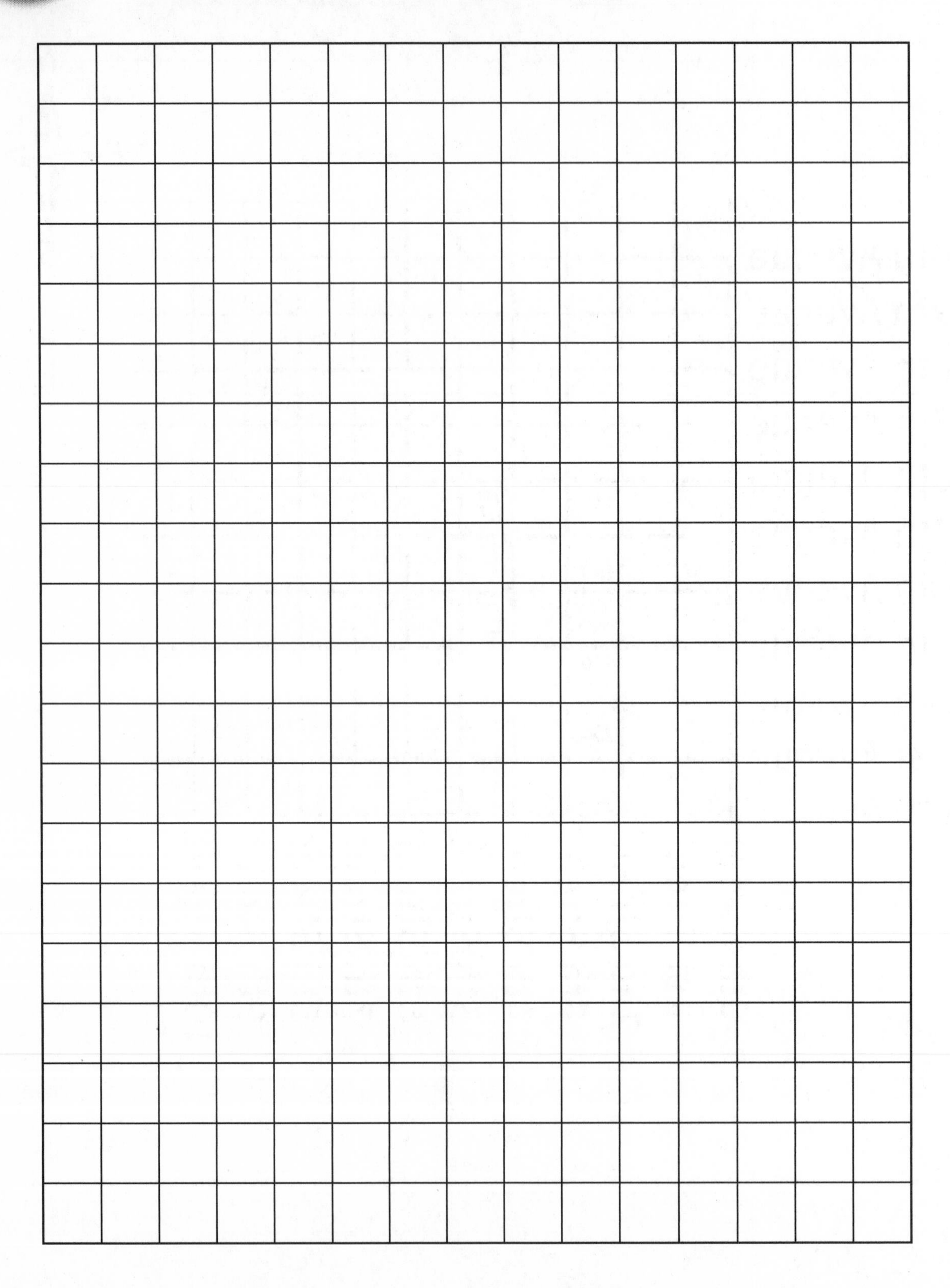